HIGH-POWER CONVERTERS AND AC DRIVES

IEEE Press
445 Hoes Lane
Piscataway, NJ 08854

IEEE Press Editorial Board
Mohamed E. El-Hawary, *Editor in Chief*

M. Akay	T. G. Croda	M. S. Newman
J. B. Anderson	R.J. Herrick	F. M. B. Pereira
R. J. Baker	S. V. Kartalopoulos	C. Singh
J. E. Brewer	M. Montrose	G. Zobrist

Kenneth Moore, *Director of IEEE Book and Information Services (BIS)*
Catherine Faduska, *Acquisitions Editor*
Jeannie Audino, *Project Editor*

HIGH-POWER CONVERTERS AND AC DRIVES

Bin Wu

GEORGE GREEN LIBRARY OF
SCIENCE AND ENGINEERING

IEEE PRESS

A John Wiley & Sons, Inc., Publication

Copyright © 2006 by the Institute of Electrical and Electronics Engineers. All rights reserved.

Published by John Wiley & Sons, Inc., Hoboken, New Jersey.
Published simultaneously in Canada.

No part of this publication may be reproduced, stored in a retrieval system, or transmitted in any form or by any means, electronic, mechanical, photocopying, recording, scanning, or otherwise, except as permitted under Section 107 or 108 of the 1976 United States Copyright Act, without either the prior written permission of the Publisher, or authorization through payment of the appropriate per-copy fee to the Copyright Clearance Center, Inc., 222 Rosewood Drive, Danvers, MA 01923, (978) 750-8400, fax (978) 750-4470, or on the web at www.copyright.com. Requests to the Publisher for permission should be addressed to the Permissions Department, John Wiley & Sons, Inc., 111 River Street, Hoboken, NJ 07030, (201) 748-6011, fax (201) 748-6008, or online at http://www.wiley.com/go/permission.

Limit of Liability/Disclaimer of Warranty: While the publisher and author have used their best efforts in preparing this book, they make no representations or warranties with respect to the accuracy or completeness of the contents of this book and specifically disclaim any implied warranties of merchantability or fitness for a particular purpose. No warranty may be created or extended by sales representatives or written sales materials. The advice and strategies contained herein may not be suitable for your situation. You should consult with a professional where appropriate. Neither the publisher nor author shall be liable for any loss of profit or any other commercial damages, including but not limited to special, incidental, consequential, or other damages.

For general information on our other products and services or for technical support, please contact our Customer Care Department within the United States at (800) 762-2974, outside the United States at (317) 572-3993 or fax (317) 572-4002.

Wiley also publishes its books in a variety of electronic formats. Some content that appears in print may not be available in electronic format. For information about Wiley products, visit our web site at www.wiley.com.

Library of Congress Cataloging-in-Publication is available.

ISBN-13 978-0-471-73171-9
ISBN-10 0-471-73171-4

Contents

Preface   xiii

Part One   Introduction   1

1. Introduction   3

    1.1   Introduction   3
    1.2   Technical Requirements and Challenges   5
         1.2.1   Line-Side Requirements   5
         1.2.2   Motor-Side Challenges   6
         1.2.3   Switching Device Constraints   7
         1.2.4   Drive System Requirements   8
    1.3   Converter Configurations   8
    1.4   MV Industrial Drives   10
    1.5   Summary   13
         References   13
         Appendix   14

2. High-Power Semiconductor Devices   17

    2.1   Introduction   17
    2.2   High-Power Switching Devices   18
         2.2.1   Diodes   18
         2.2.2   Silicon-Controlled Rectifier (SCR)   18
         2.2.3   Gate Turn-Off (GTO) Thyristor   21
         2.2.4   Gate-Commutated Thyristor (GCT)   23
         2.2.5   Insulated Gate Bipolar Transistor (IGBT)   26
         2.2.6   Other Switching Devices   28
    2.3   Operation of Series-Connected Devices   28
         2.3.1   Main Causes of Voltage Unbalance   29
         2.3.2   Voltage Equalization for GCTs   29
         2.3.3   Voltage Equalization for IGBTs   31
    2.4   Summary   32
         References   33

Part Two Multipulse Diode and SCR Rectifiers   35

3. Multipulse Diode Rectifiers   37

- 3.1 Introduction   37
- 3.2 Six-Pulse Diode Rectifier   38
  - 3.2.1 Introduction   38
  - 3.2.2 Capacitive Load   40
  - 3.2.3 Definition of THD and PF   43
  - 3.2.4 Per-Unit System   45
  - 3.2.5 THD and PF of Six-Pulse Diode Rectifier   45
- 3.3 Series-Type Multipulse Diode Rectifiers   47
  - 3.3.1 12-Pulse Series-Type Diode Rectifier   47
  - 3.3.2 18-Pulse Series-Type Diode Rectifier   51
  - 3.3.3 24-Pulse Series-Type Diode Rectifier   54
- 3.4 Separate-Type Multipulse Diode Rectifiers   57
  - 3.4.1 12-Pulse Separate-Type Diode Rectifier   57
  - 3.4.2 18- and 24-Pulse Separate-Type Diode Rectifiers   61
- 3.5 Summary   61
- References   61

4. Multipulse SCR Rectifiers   65

- 4.1 Introduction   65
- 4.2 Six-Pulse SCR Rectifier   65
  - 4.2.1 Idealized Six-Pulse Rectifier   66
  - 4.2.2 Effect of Line Inductance   70
  - 4.2.3 Power Factor and THD   72
- 4.3 12-Pulse SCR Rectifier   74
  - 4.3.1 Idealized 12-Pulse Rectifier   75
  - 4.3.2 Effect of Line and Leakage Inductances   78
  - 4.3.3 THD and PF   79
- 4.4 18- and 24-Pulse SCR Rectifiers   79
- 4.5 Summary   81
- References   81

5. Phase-Shifting Transformers   83

- 5.1 Introduction   83
- 5.2 Y/Z Phase-Shifting Transformers   83
  - 5.2.1 Y/Z-1 Transformers   83
  - 5.2.2 Y/Z-2 Transformers   85
- 5.3 Δ/Z Transformers   87
- 5.4 Harmonic Current Cancellation   88
  - 5.4.1 Phase Displacement of Harmonic Currents   88
  - 5.4.2 Harmonic Cancellation   90
- 5.5 Summary   92

Contents vii

Part Three   Multilevel Voltage Source Converters   93

6. Two-Level Voltage Source Inverter   95

    6.1   Introduction   95
    6.2   Sinusoidal PWM   95
        6.2.1   Modulation Scheme   95
        6.2.2   Harmonic Content   96
        6.2.3   Overmodulation   99
        6.2.4   Third Harmonic Injection PWM   99
    6.3   Space Vector Modulation   101
        6.3.1   Switching States   101
        6.3.2   Space Vectors   101
        6.3.3   Dwell Time Calculation   104
        6.3.4   Modulation Index   106
        6.3.5   Switching Sequence   107
        6.3.6   Spectrum Analysis   108
        6.3.7   Even-Order Harmonic Elimination   111
        6.3.8   Discontinuous Space Vector Modulation   115
    6.4   Summary   116
        References   117

7. Cascaded H-Bridge Multilevel Inverters   119

    7.1   Introduction   119
    7.2   H-Bridge Inverter   119
        7.2.1   Bipolar Pulse-Width Modulation   120
        7.2.2   Unipolar Pulse-Width Modulation   121
    7.3   Multilevel Inverter Topologies   123
        7.3.1   CHB Inverter with Equal dc Voltage   123
        7.3.2   H-Bridges with Unequal dc Voltages   126
    7.4   Carrier Based PWM Schemes   127
        7.4.1   Phase-Shifted Multicarrier Modulation   127
        7.4.2   Level-Shifted Multicarrier Modulation   131
        7.4.3   Comparison Between Phase- and Level-Shifted
            PWM Schemes   136
    7.5   Staircase Modulation   139
    7.6   Summary   141
        References   142

8. Diode-Clamped Multilevel Inverters   143

    8.1   Introduction   143
    8.2   Three-Level Inverter   143
        8.2.1   Converter Configuration   143
        8.2.2   Switching State   144

        8.2.3   Commutation   145
  8.3   Space Vector Modulation   148
        8.3.1   Stationary Space Vectors   149
        8.3.2   Dwell Time Calculation   149
        8.3.3   Relationship Between $\vec{V}_{ref}$ Location and Dwell Times   154
        8.3.4   Switching Sequence Design   154
        8.3.5   Inverter Output Waveforms and Harmonic Content   160
        8.3.6   Even-Order Harmonic Elimination   160
  8.4   Neutral-Point Voltage Control   164
        8.4.1   Causes of Neutral-Point Voltage Deviation   165
        8.4.2   Effect of Motoring and Regenerative Operation   165
        8.4.3   Feedback Control of Neutral-Point Voltage   166
  8.5   Other Space Vector Modulation Algorithms   167
        8.5.1   Discontinuous Space Vector Modulation   167
        8.5.2   SVM Based on Two-Level Algorithm   168
  8.6   High-Level Diode-Clamped Inverters   168
        8.6.1   Four- and Five-Level Diode-Clamped Inverters   169
        8.6.2   Carrier-Based PWM   170
  8.7   Summary   173
       References   175
       Appendix   176

9. Other Multilevel Voltage Source Inverters   179

  9.1   Introduction   179
  9.2   NPC/H-Bridge Inverter   179
        9.2.1   Inverter Topology   179
        9.2.2   Modulation Scheme   180
        9.2.3   Waveforms and Harmonic Content   181
  9.3   Multilevel Flying-Capacitor Inverters   183
        9.3.1   Inverter Configuration   183
        9.3.2   Modulation Schemes   184
  9.4   Summary   186
       References   186

Part Four   PWM Current Source Converters   187

10. PWM Current Source Inverters   189

  10.1   Introduction   189
  10.2   PWM Current Source Inverter   190
        10.2.1   Trapezoidal Modulation   191
        10.2.2   Selective Harmonic Elimination   194

10.3 Space Vector Modulation   200
    10.3.1 Switching States   200
    10.3.2 Space Vectors   201
    10.3.3 Dwell Time Calculation   203
    10.3.4 Switching Sequence   205
    10.3.5 Harmonic Content   208
    10.3.6 SVM Versus TPWM and SHE   209
10.4 Parallel Current Source Inverters   209
    10.4.1 Inverter Topology   209
    10.4.2 Space Vector Modulation for Parallel Inverters   210
    10.4.3 Effect of Medium Vectors on dc Currents   212
    10.4.4 dc Current Balance Control   213
    10.4.5 Experimental Verification   214
10.5 Load-Commutated Inverter (LCI)   215
10.6 Summary   216
    References   217
    Appendix   218

11. PWM Current Source Rectifiers   219

11.1 Introduction   219
11.2 Single-Bridge Current Source Rectifier   219
    11.2.1 Introduction   219
    11.2.2 Selective Harmonic Elimination   220
    11.2.3 Rectifier dc Output Voltage   225
    11.2.4 Space Vector Modulation   227
11.3 Dual-Bridge Current Source Rectifier   227
    11.3.1 Introduction   227
    11.3.2 PWM Schemes   228
    11.3.3 Harmonic Contents   229
11.4 Power Factor Control   231
    11.4.1 Introduction   231
    11.4.2 Simultaneous α and m_a Control   232
    11.4.3 Power Factor Profile   235
11.5 Active Damping Control   236
    11.5.1 Introduction   236
    11.5.2 Series and Parallel Resonant Modes   237
    11.5.3 Principle of Active Damping   238
    11.5.4 LC Resonance Suppression   240
    11.5.5 Harmonic Reduction   242
    11.5.6 Selection of Active Damping Resistance   245
11.6 Summary   246
    References   247
    Appendix   248

Part Five High-Power AC Drives — 251

12. Voltage Source Inverter-Fed Drives — 253

12.1 Introduction 253
12.2 Two-Level VBSI-Based MV Drives 253
    12.2.1 Power Converter Building Block 253
    12.2.2 Two-Level VSI with Passive Front End 254
12.3 Neutral-Point Clamped (NPC) Inverter-Fed Drives 257
    12.3.1 GCT-Based NPC Inverter Drives 257
    12.3.2 IGBT-Based NPC Inverter Drives 260
12.4 Multilevel Cascaded H-Bridge (CHB) Inverter-Fed Drives 261
    12.4.1 CHB Inverter-Fed Drives for 2300-V/4160-V Motors 261
    12.4.2 CHB Inverter Drives for 6.6-kV/11.8-kV Motors 264
12.5 NPC/H-Bridge Inverter-Fed Drives 264
12.6 Summary 265
    References 265

13. Current Source Inverter-Fed Drives — 269

13.1 Introduction 269
13.2 CSI Drives with PWM Rectifiers 269
    13.2.1 CSI Drives with Single-bridge PWM Rectifier 269
    13.2.2 CSI Drives for Custom Motors 273
    13.2.3 CSI Drives with Dual-Bridge PWM Rectifier 275
13.3 Transformerless CSI Drive for Standard AC Motors 276
    13.3.1 CSI Drive Configuration 276
    13.3.2 Integrated dc Choke for Common-Mode Voltage Suppression 277
13.4 CSI Drive with Multipulse SCR Rectifier 279
    13.4.1 CSI Drive with 18-Pulse SCR Rectifier 279
    13.4.2 Low-Cost CSI Drive with 6-Pulse SCR Rectifier 280
13.5 LCI Drives for Synchronous Motors 281
    13.5.1 LCI Drives with 12-Pulse Input and 6-Pulse Output 281
    13.5.2 LCI Drives with 12-Pulse Input and 12-Pulse Output 282
13.6 Summary 282
    References 283

14. Advanced Drive Control Schemes — 285

14.1 Introduction 285
14.2 Reference Frame Transformation 285
    14.2.1 abc/dq Frame Transformation 286
    14.2.2 3/2 Stationary Transformation 288
14.3 Induction Motor Dynamic Models 288
    14.3.1 Space Vector Motor Model 288

        14.3.2 dq-Axis Motor Model   290
        14.3.3 Induction Motor Transient Characteristics   291
  14.4 Principle of Field-Oriented Control (FOC)   296
        14.4.1 Field Orientation   296
        14.4.2 General Block Diagram of FOC   297
  14.5 Direct Field-Oriented Control   298
        14.5.1 System Block Diagram   298
        14.5.2 Rotor Flux Calculator   299
        14.5.3 Direct FOC with Current-Controlled VSI   301
  14.6 Indirect Field-Oriented Control   305
  14.7 FOC for CSI-Fed Drives   307
  14.8 Direct Torque Control   309
        14.8.1 Principle of Direct Torque Control   310
        14.8.2 Switching Logic   311
        14.8.3 Stator Flux and Torque Calculation   313
        14.8.4 DTC Drive Simulation   314
        14.8.5 Comparison Between DTC and FOC Schemes   316
  14.9 Summary   316
        References   317

Abbreviations   319

Appendix   Projects for Graduate-Level Courses   321

  P. 1 Introduction   321
  P. 2 Sample Project   322
  P. 3 Answers to Sample Project   324

Index   329

About the Author   333

Preface

With technology advancements in semiconductor devices such as insulated gate bipolar transistors (IGBTs) and gate commutated thyristors (GCTs), modern high-power medium voltage (MV) drives are increasingly used in petrochemical, mining, steel and metals, transportation and other industries to conserve electric energy, increase productivity and improve product quality.

Although research and development of the medium voltage (2.3 KV to 13.8 KV) drive in the 1-MW to 100-MW range are continuously growing, books dedicated to this technology seem unavailable. This book provides a comprehensive analysis on a variety of high-power converter topologies, drive system configurations, and advanced control schemes.

This book presents the latest technology in the field, provides design guidance with tables, charts and graphs, addresses practical problems and their mitigation methods, and illustrates important concepts with computer simulations and experiments. It serves as a reference for academic researchers, practicing engineers, and other professionals. This book also provides adequate technical background for most of its topics such that it can be adopted as a textbook for a graduate-level course in power electronics and ac drives.

This book is presented in five parts with fourteen chapters. Part One, Introduction, provides an overview of high-power MV drives, which includes market analysis, drive system configurations, typical industrial applications, power converter topologies and semiconductor devices. The technical requirements and challenges for the MV drive are highlighted; these are different in many aspects from those for low-voltage drives.

Part Two, Multipulse Diode and SCR Rectifiers, covers 12-, 18- and 24-pulses rectifier topologies commonly used in the MV drive for the reduction of line current distortion. The configuration of phase-shifting (zigzag) transformers and principle of harmonic cancellation are discussed.

Part Three, Multilevel Voltage Source Inverters, presents detailed analysis on various multilevel voltage source inverter (VSI) topologies, including neutral point clamped and cascaded H-bridge inverters. Carrier-based and space-vector modulation schemes for the multilevel inverters are elaborated.

Part Four, PWM Current Source Converters, deals with a number of current source inverters (CSI) and rectifiers for the MV drive. Several modulation techniques such as trapezoidal pulse width modulations, selective harmonics elimination and space vector modulations are analyzed. Unity-power factor control and active damping control for the current source rectifiers are also included.

Part Five, High-Power ac Drives, focuses on various configurations of VSI- and CSI-fed MV drives marketed by major drive manufacturers. The features and limitations of these drives are discussed. Two advanced drive control schemes, field oriented control and direct torque control, are analyzed. Efforts are made to present these complex schemes in a simple, easy to understand manner.

The Appendix at the end of the book provides a list of 12 simulation based projects for use in a graduate course. The detailed instruction for the projects and their answers are included in Instructor's Manual (published separately). Since the book is rich in illustrations, Power Point slides for each of the chapters are included in the manual.

Finally, I would like to express my deep gratitude to my colleagues at Rockwell Automation Canada; in particular, Steve Rizzo, Navid Zargari, and Frank DeWinter, for numerous discussions and 12 years of working together in developing advanced MV-drive technologies. I sincerely thank my supervisors, Drs. Shashi Dowan and Gordon Slemon for their valuable advice on high-power drive research during my graduate studies at the University of Toronto. I am also indebted to Dr. Robert Hanna at RPM Engineering Ltd. for his review of the manuscript and constructive comments. I am grateful to my postdoctoral fellows and graduate students in the Laboratory for Electric Drive Applications and Research (LEDAR) at Ryerson University for their assistance in preparing the manuscript of this book. I am thankful to my colleagues at ASI Robicon, ABB, Siemens AG, and Rockwell Automation for providing the photos of the MV drives. I also wish to acknowledge the support and inspiration of my wife, Janice, and my daughter, Linda, during the preparation of this book.

BIN WU

Toronto, Canada
December 2005

Part One

Introduction

Chapter 1

Introduction

1.1 INTRODUCTION

The development of high-power converters and medium-voltage (MV) drives started in the mid-1980s when 4500-V gate turn off (GTO) thyristors became commercially available [1]. The GTO was the standard for the MV drive until the advent of high-power insulated gate bipolar transistors (IGBTs) and gate commutated thyristors (GCTs) in the late 1990s [2, 3]. These switching devices have rapidly progressed into the main areas of high-power electronics due to their superior switching characteristics, reduced power losses, ease of gate control, and snubberless operation.

The MV drives cover power ratings from 0.4 MW to 40 MW at the medium-voltage level of 2.3 kV to 13.8 kV. The power rating can be extended to 100 MW, where synchronous motor drives with load commutated inverters are often used [4]. However, the majority of the installed MV drives are in the 1- to 4-MW range with voltage ratings from 3.3 kV to 6.6 kV as illustrated in Fig. 1.1-1.

The high-power MV drives have found widespread applications in industry. They are used for pipeline pumps in the petrochemical industry [5], fans in the cement industry [6], pumps in water pumping stations [7], traction applications in the transportation industry [8], steel rolling mills in the metals industry [9], and other applications [10,11]. A summary of the MV drive applications is given in the appendix of this chapter [12].

Since the beginning of the 21st century a few thousands of MV drives have been commissioned worldwide. Market research has shown that around 85% of the total installed drives are for pumps, fans, compressors and conveyors [13], where the drive system may not require high dynamic performance. As shown in Fig. 1.1-2, only 15% of the installed drives are nonstandard drives.

One of the major markets for the MV drive is for retrofit applications. It is reported that 97% of the currently installed MV motors operate at a fixed speed and only 3% of them are controlled by variable-speed drives [13]. When fans or pumps are driven by a fixed-speed motor, the control of air or liquid flow is normally achieved by conventional mechanical methods, such as throttling control, inlet dampers, and flow control valves, resulting in a substantial amount of energy loss.

4 Chapter 1 Introduction

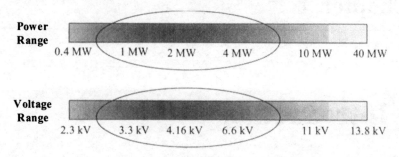

Figure 1.1-1 Voltage and power ranges of the MV drive. Source: Rockwell Automation.

The installation of the MV drive can lead to a significant savings on energy cost. It was reported that the use of the variable-speed MV drive resulted in a payback time of the investment from one to two and a half years [7].

The use of the MV drive can also increase productivity in some applications. A case was reported from a cement plant where the speed of a large fan was made adjustable by an MV drive [11]. The collected dust on the fan blades operated at a fixed speed had to be cleaned regularly, leading to a significant downtime per year for maintenance. With variable-speed operation, the blades only had to be cleaned at the standstill of the production once a year. The increase in productivity together with the energy savings resulted in a payback time of the investment within six months.

Figure 1.1-3 shows a general block diagram of the MV drive. Depending on the system requirements and the type of the converters employed, the line- and motor-side filters are optional. A phase shifting transformer with multiple secondary windings is often used mainly for the reduction of line current distortion.

The rectifier converts the utility supply voltage to a dc voltage with a fixed or adjustable magnitude. The commonly used rectifier topologies include multipulse

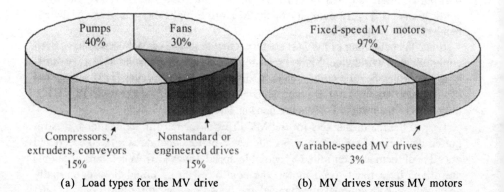

Figure 1.1-2 MV drive market survey. Source: ABB.

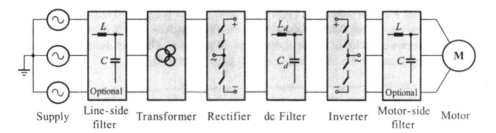

Figure 1.1-3 General block diagram of the MV drive.

diode rectifiers, multipulse SCR rectifiers, or pulse-width-modulated (PWM) rectifiers. The dc filter can simply be a capacitor that provides a stiff dc voltage in voltage source drives or an inductor that smoothes the dc current in current source drives.

The inverter can be generally classified into voltage source inverter (VSI) and current source inverter (CSI). The VSI converts the dc voltage to a three-phase ac voltage with adjustable magnitude and frequency whereas the CSI converts the dc current to an adjustable three-phase ac current. A variety of inverter topologies have been developed for the MV drive, most of which will be analyzed in this book.

1.2 TECHNICAL REQUIREMENTS AND CHALLENGES

The technical requirements and challenges for the MV drive differ in many aspects from those for the low-voltage ($\leq$ 600 V) ac drives. Some of them that must be addressed in the MV drive may not even be an issue for the low-voltage drives. These requirements and challenges can be generally divided into four groups: the requirements related to the power quality of line-side converters, the challenges associated with the design of motor-side converters, the constraints of the switching devices, and the drive system requirements.

1.2.1 Line-Side Requirements

(a) Line Current Distortion. The rectifier normally draws distorted line current from the utility supply, and it also causes notches in voltage waveforms. The distorted current and voltage waveforms can cause numerous problems such as nuisance tripping of computer-controlled industrial processes, overheating of transformers, equipment failure, computer data loss, and malfunction of communications equipment. Nuisance tripping of industrial assembly lines often leads to expensive downtime and ruined product. There exist certain guidelines for harmonic regulation, such as IEEE Standard 519-1992 [14]. The rectifier used in the MV drive should comply with these guidelines.

(b) Input Power Factor. High input power factor is a general requirement for all electric equipment. Most of the electric utility companies require their customers to have a power factor of 0.9 or above to avoid penalties. This requirement is especially important for the MV drive due to its high power rating.

(c) LC Resonance Suppression. For the MV drives using line-side capacitors for current THD reduction or power factor compensation, the capacitors form LC resonant circuits with the line inductance of the system. The LC resonant modes may be excited by the harmonic voltages in the utility supply or harmonic currents produced by the rectifier. Since the utility supply at the medium voltage level normally has very low line resistance, the lightly damped LC resonances may cause severe oscillations or overvoltages that may destroy the switching devices and other components in the rectifier circuits. The LC resonance issue should be addressed when the drive system is designed.

1.2.2 Motor-Side Challenges

(a) dv/dt and Wave Reflections. Fast switching speed of the semiconductor devices results in high dv/dt at the rising and falling edges of the inverter output voltage waveform. Depending on the magnitude of the inverter dc bus voltage and speed of the switching device, the dv/dt can well exceed 10,000 V/μs. The high dv/dt in the inverter output voltage can cause premature failure of the motor winding insulation due to partial discharges. It induces rotor shaft voltages through stray capacitances between the stator and rotor. The shaft voltage produces a current flowing into the shaft bearing, leading to early bearing failure. The high dv/dt also causes electromagnetic emission in the cables connecting the motor to the inverter, affecting the operation of nearby sensitive electronic equipment.

To make the matter worse, the high dv/dt may cause a voltage doubling effect at the rising and falling edges of the motor voltage waveform due to wave reflections in long cables. The reflections are caused by the mismatch between the wave impedance of the cable and the impedances at its inverter and motor ends, and they can double the voltage on the motor terminals at each switching transient if the cable length exceeds a certain limit. The critical cable length for 500 V/μs is in the 100-m range, for 1000 V/μs in the 50-m range, and for 10,000 V/μs in the 5-m range [15].

(b) Common-Mode Voltage Stress. The switching action of the rectifier and inverter normally generates common-mode voltages [16]. The common-mode voltages are essentially zero-sequence voltages superimposed with switching noise. If not mitigated, they will appear on the neutral of the stator winding with respect to ground, which should be zero when the motor is powered by a three-phase balanced utility supply. Furthermore, the motor line-to-ground voltage, which should be equal to the motor line-to-neutral (phase) voltage, can be substantially increased

due to the common-mode voltages, leading to the premature failure of the motor winding insulation system. As a consequence, the motor life expectancy is shortened.

It is worth noting that the common-mode voltages are generated by the rectification and inversion process of the converters. This phenomenon is different from the high dv/dt caused by the switching transients of the high speed switches. It should be further noted that the common-mode voltage issue is often ignored in the low-voltage drives. This is partially due to the conservative design of the insulation system for low-voltage motors. In the MV drives, the motor should not be subject to any common-mode voltages. Otherwise, the replacement of the damaged motor would be very costly in addition to the loss of production.

(c) Motor Derating. High-power inverters may generate a large amount of current and voltage harmonics. These harmonics cause additional power losses in the motor winding and magnetic core. As a consequence, the motor is derated and cannot operate at its full capacity.

(d) LC Resonances. For the MV drives with a motor-side filter capacitor, the capacitor forms an LC resonant circuit with the motor inductances. The resonant mode of the LC circuit may be excited by the harmonic voltages or currents produced by the inverter. Although the motor winding resistances may provide some damping, this problem should be addressed at the design stage of the drive.

(e) Torsional Vibration. Torsional vibrations may occur in the MV drive due to the large inertias of the motor and its mechanical load. The drive system may vary from a simple two-inertia system consisting of only the motor and the load inertias to very complex systems such as a steel rolling-mill drive with more than 20 inertias. The torsional vibrations may be excited when the natural frequency of the mechanical system is coincident with the frequency of torque pulsations caused by distorted motor currents. Excessive torsional vibrations can result in broken shafts and couplings, and also cause damages to the other mechanical components in the system.

1.2.3 Switching Device Constraints

(a) Device Switching Frequency. The device switching loss accounts for a significant amount of the total power loss in the MV drive. The switching loss minimization can lead to a reduction in the operating cost when the drive is commissioned. The physical size and manufacturing cost of the drive can also be reduced due to the reduced cooling requirements for the switching devices. The other reason for limiting the switching frequency is related to the device thermal resistance that may prevent efficient heat transfer from the device to its heatsink. In practice, the device switching frequency is normally around 200 Hz for GTOs and 500 Hz for IGBTs and GCTs.

The reduction of switching frequency generally causes an increase in harmonic distortion of the line- and motor-side waveforms of the drive. Efforts should be made to minimize the waveform distortion with limited switching frequencies.

(b) Series Connection. Switching devices in the MV drive are often connected in series for medium-voltage operation. Since the series connected devices and their gate drivers may do not have identical static and dynamic characteristics, they may not equally share the total voltage in the blocking mode or during switching transients. A reliable voltage equalization scheme should be implemented to protect the switching devices and enhance the system reliability.

1.2.4 Drive System Requirements

The general requirements for the MV drive system include high efficiency, low manufacturing cost, small physical size, high reliability, effective fault protection, easy installation, self-commissioning, and minimum downtime for repairs. Some of the application-specific requirements include high dynamic performance, regenerative braking capability, and four-quadrant operation.

1.3 CONVERTER CONFIGURATIONS

Multipulse rectifiers are often employed in the MV drive to meet the line-side harmonic requirements. Figure 1.3-1 illustrates a block diagram of 12-, 18- and 24-pulse rectifiers. Each multipulse rectifier is essentially composed of a phase-shifting transformer with multiple secondary windings feeding a set of identical six-pulse rectifiers.

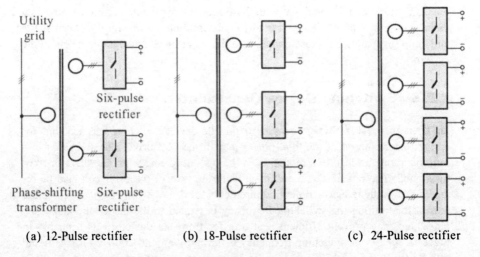

Figure 1.3-1 Multipulse diode/SCR rectifiers.

1.3 Converter Configurations   9

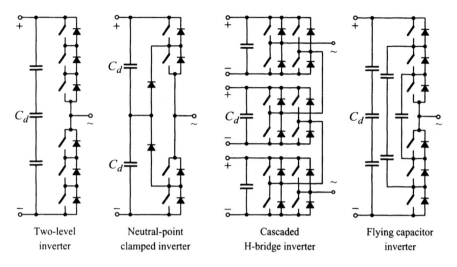

| Two-level inverter | Neutral-point clamped inverter | Cascaded H-bridge inverter | Flying capacitor inverter |

Figure 1.3-2 Per-phase diagram of VSI topologies.

Both diode and SCR devices can be used as switching devices. The multipulse diode rectifiers are suitable for VSI-fed drives while the SCR rectifiers are normally for CSI drives. Depending on the inverter configuration, the outputs of the six-pulse rectifiers can be either connected in series to form a single dc supply or connected directly to a multilevel inverter that requires isolated dc supplies. In addition to the diode and SCR rectifiers, PWM rectifiers using IGBT or GCT devices can also be employed, where the rectifier usually has the same topology as the inverter.

To meet the motor-side challenges, a variety of inverter topologies can be adopted for the MV drive. Figure 1.3-2 illustrates per-phase diagram of commonly used

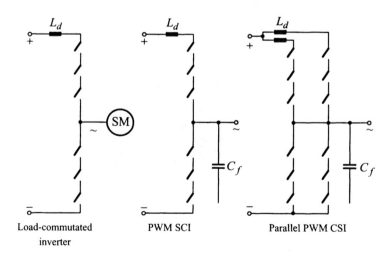

Load-commutated inverter   PWM SCI   Parallel PWM CSI

Figure 1.3-3 Per-phase diagram of CSI topologies.

three-phase multilevel VSI topologies, which include a conventional two-level inverter, a three-level neutral-point clamped (NPC) inverter, a seven-level cascaded H-bridge inverter and a four-level flying-capacitor inverter. Either IGBT or GCT can be employed in these inverters as a switching device.

Current source inverter technology has been widely accepted in the drive industry. Figure 1.3-3 shows the per-phase diagram of the CSI topologies for the MV drive. The SCR-based load-commutated inverter (LCI) is specially suitable for very large synchronous motor drives, while the PWM current source inverter is a preferred choice for most industrial applications. The parallel PWM CSI is composed of two or more single-bridge inverters connected in parallel for super-high-power applications. Symmetrical GCTs are normally used in the PWM current source inverters.

1.4 MV INDUSTRIAL DRIVES

A number of MV drive products are available on the market today. These drives come with different designs using various power converter topologies and control schemes. Each design offers some unique features but also has some limitations. The diversified offering promotes the advancement in the drive technology and the market competition as well. A few examples of the MV industrial drives are as follows.

Figure 1.4-1 illustrates the picture of an MV drive rated at 4.16 kV and 1.2 MW. The drive is composed of a 12-pulse diode rectifier as a front end and a three-level

Figure 1.4-1 GCT-based three-level NPC inverter-fed MV drive. Courtesy of ABB (ACS1000).

1.4 MV Industrial Drives 11

Figure 1.4-2 IGBT-based three-level NPC inverter-fed MV drive. Courtesy of Siemens (SIMOVERT MV).

NPC inverter using GCT devices. The drive's digital controller is installed in the left cabinet. The cabinet in the center houses the diode rectifier and air-cooling system of the drive. The inverter and its output filters are mounted in the right cabinet. The phase-shifting transformer for the rectifier is normally installed outside the drive cabinets.

Figure 1.4-2 shows an MV drive using an IGBT-based three-level NPC inverter. The IGBT–heatsink assemblies in the central cabinet are constructed in a modular fashion for easy assembly and replacement. The front end converter is a standard 12-pulse diode rectifier for line current harmonic reduction. The phase-shifting transformer for the rectifier is not included in the drive cabinet.

A 4.16-kV 7.5-MW cascaded H-bridge inverter-fed drive is illustrated in Fig. 1.4-3. The inverter is composed of 15 identical IGBT power cells, each of which can be slid out for quick repair or replacement. The waveform of the inverter line-to-line voltage is composed of 21 levels, leading to near-sinusoidal waveforms without using LC filters. The drive employs a 30-pulse diode rectifier powered by a phase-shifting transformer with 15 secondary windings. The transformer is installed in the left cabinets to reduce the installation cost of the cables connecting its secondary windings to the power cells.

Figure 1.4-4 shows a current source inverter-fed MV drive with a power range from 2.3 MW to 7 MW. The drive comprises two identical PWM GCT current

12 Chapter 1 Introduction

Figure 1.4-3 IGBT cascaded H-bridge inverter-fed MV drive. Courtesy of ASI Robicon (Perfect Harmony).

Figure 1.4-4 CSI-fed MV drive using symmetrical GCTs. Courtesy of Rockwell Automation (PowerFlex 7000).

Table 1.4-1 Summary of the MV Drive Products Marketed by Major Drive Manufacturers

Inverter Configuration	Switching Device	Power Range (MVA)	Manufacturer
Two-level voltage source inverter	IGBT	1.4–7.2	Alstom (VDM5000)
Three-level neutral point clamped inverter	GCT	0.3–5 3–27	ABB (ACS1000) (ACS6000)
	GCT	3–20	General Electric (Innovation Series MV-SP)
	IGBT	0.6–7.2	Siemens (SIMOVERT-MV)
	IGBT	0.3–2.4	General Electric-Toshiba (Dura-Bilt5 MV)
Multilevel cascaded H-bridge inverter	IGBT	0.3–22	ASI Robicon (Perfect Harmony)
		0.5–6	Toshiba (TOSVERT-MV)
		0.45–7.5	General Electric (Innovation MV-GP Type H)
NPC/H-bridge inverter	IGBT	0.4–4.8	Toshiba (TOSVERT 300 MV)
Flying-capacitor inverter	IGBT	0.3–8	Alstom (VDM6000 Symphony)
PWM current source inverter	Symmetrical GCT	0.2–20	Rockwell Automation (PowerFlex 7000)
Load commutated inverter	SCR	>10 >10 >10	Siemens (SIMOVERT S) ABB (LCI) Alstom (ALSPA SD7000)

source converters, one for the rectifier and the other for the inverter. The converters are installed in the second cabinet from the left. The dc inductor required by the current source drive is mounted in the fourth cabinet. The fifth (right most) cabinet contains drive's liquid cooling system. With the use of a special integrated dc inductor having both differential- and common-mode inductances, the drive does not require an isolation transformer for the common-mode voltage mitigation, leading to a reduction in manufacturing cost.

Table 1.4-1 provides a summary of the MV drive products offered by major drive manufacturers in the world, where the inverter configuration, switching device, and power range of the drive are listed.

1.5 SUMMARY

This chapter provides an overview of high-power converters and medium-voltage (MV) drives, including market analysis, drive system configurations, power converter topologies, drive product analysis, and major manufacturers. The technical requirements and challenges for the MV drive are also summarized. These requirements and challenges will be addressed in the subsequent chapters, where various power converters and MV drive systems are analyzed.

REFERENCES

1. S. Rizzo and N. Zargari, Medium Voltage Drives: What Does the Future Hold? *The 4th International Power Electronics and Motional Control Conference (IPEMC)*, pp. 82–89, 2004.
2. H. Brunner, M. Hieholzer, et al., Progress in Development of the 3.5 kV High Voltage IGBT/Diode Chipset and 1200A Module Applications, *IEEE International Symposium on Power Semiconductor Devices and IC's*, pp. 225–228, 1997.
3. P. K. Steimer, H. E. Gruning, et al., IGCT—A New Emerging Technology for High Power Low Cost Inverters, *IEEE Industry Application Magazine*, pp. 12–18, 1999.
4. R. Bhatia, H. U. Krattiger, A. Bonanini, et al., Adjustable Speed Drive with a Single 100-MW Synchronous Motor, *ABB Review*, No. 6, pp. 14–20, 1998.
5. W. C. Rossmann and R. G. Ellis, Retrofit of 22 Pipeline Pumping Stations with 3000-hp Motors and Variable-Frequency Drives, *IEEE Transactions on Industry Applications*, Vol. 34, Issue: 1, pp. 178–186, 1998.
6. R. Menz and F. Opprecht, Replacement of a Wound Rotor Motor with an Adjustable Speed Drive for a 1400 kW Kiln Exhaust Gas Fan, *The 44th IEEE IAS Cement Industry Technical Conference*, pp. 85–93, 2002.
7. B. P. Schmitt and R. Sommer, Retrofit of Fixed Speed Induction Motors with Medium Voltage Drive Converters Using NPC Three-Level Inverter High-Voltage IGBT Based Topology, *IEEE International Symposium on Industrial Electronics*, pp. 746–751, 2001.
8. S. Bernert, Recent Development of High Power Converters for Industry and Traction Applications, *IEEE Transactions on Power Electronics*, Vol. 15, No. 6, pp. 1102–1117, 2000.
9. H. Okayama, M. Koyama, et al., Large Capacity High Performance 3-level GTO Inverter System for Steel Main Rolling Mill Drives, *IEEE Industry Application Society (IAS) Conference*, pp. 174–179, 1996.
10. N. Akagi, Large Static Converters for Industry and Utility Applications, *IEEE Proceedings*, Vol. 89, No. 6, pp. 976–983, 2001.
11. R. A. Hanna and S. Randall, Medium Voltage Adjustable Speed Drive Retrofit of an Existing Eddy Current Clutch Extruder Application, *IEEE Transaction on Industry Applications*, Vol. 33, No. 6, pp. 1750–1755.
12. N. Zargari and S. Rizzo, Medium Voltage Drives in Industrial Applications, Technical Seminar, IEEE Toronto Section, 37 pages, November 2004.
13. S. Malik and D. Kluge, ACS1000 World's First Standard AC Drive for Medium-Voltage Applications, *ABB Review*, No. 2, pp. 4–11, 1998.
14. IEEE Standard 519-1992, *IEEE Recommended Practices and Requirements for Harmonic Control In Electrical Power Systems*, IEEE Inc.,1993.
15. J. K. Steinke, Use of an LC Filter to Achieve a Motor-Friendly Performance of the PWM Voltage Source Inverter, *IEEE Transactions on Energy Conversion*, Vol. 14, No. pp. 649–654, 1999.
16. S. Wei, N. Zargari, B. Wu, et al., Comparison and Mitigation of Common Mode Volt-

ages in Power Converter Topologies, *IEEE Industry Application Society (IAS) Conference*, pp. 1852–1857, 2004.

APPENDIX

A Summary of MV Drive Applications

Industry	Application Examples
Petrochemical	Pipeline pumps, gas compressors, brine pumps, mixers/extruders, electrical submersible pumps, induced draft fans, boiler feed water pumps, water injection pumps.
Cement	Kiln-induced draft fans, forced draft fans, baghouse fans, preheat tower fans, raw mill induced draft fans, kiln gas fans, cooler exhaust fans, separator fans.
Mining and Metals	Slurry pumps, ventilation fans, de-scaling pumps, tandem belt conveyors, baghouse fans, cyclone feed pumps, crushers, rolling mills, hoists, coilers, winders.
Water/Wastewater	Raw sewage pumps, bio-roughing tower pumps, treatment pumps, freshwater pumps, storm water pumps.
Transportation	Propulsion for naval vessels, shuttle tankers, icebreakers, cruisers. Traction drives for locomotives, light-track trains.
Electric Power	Feed water pumps, induced draft fans, forced draft fans, effluent pumps, compressors.
Forest Products	Induced draft fans, boiler-feed water pumps, pulpers, refiners, kiln drives, line shafts.
Miscellaneous	Wind tunnels, agitators, test stands, rubber mixers.

Chapter 2

High-Power Semiconductor Devices

2.1 INTRODUCTION

The development of semiconductor switching devices is essentially a search for the ideal switch. The effort has been made to reduce device power losses, increase switching frequencies, and simplify gate drive circuits. The evolution of the switching devices leads the pace of high-power converter development, and in the meantime the wide application of the high-power converters in industry drives the semiconductor technology toward higher power ratings with improved reliability and reduced cost.

There are two major types of high-power switching devices for use in various converters: the thyristor- and transistor-based devices. The former includes silicon-controlled rectifier (SCR), gate turn-off thyristor (GTO), and gate commutated thyristor (GCT), while the latter embraces insulated gate bipolar transistor (IGBT) and injection-enhanced gate transistor (IEGT). Other devices such as power MOSFET, emitter turn-off thyristor (ETO), MOS-controlled thyristor (MCT), and static induction thyristor (SIT) have not gained significant importance in high-power applications.

Figure 2.1-1 shows the voltage and current ratings of major switching devices commercially available for high-power converters [1]. Semiconductor manufacturers can offer SCRs rated at 12 kV/1.5 kA or 4.8 kV/5 kA. The GTO and GCT devices can reach the voltage and current ratings of 6 kV and 6 kA. The ratings of IGBT devices are relatively low, but can reach as high as 6.5 kV/0.6 kA or 1.7 kV/3.6 kA.

In this chapter the characteristics of commonly used high-power semiconductor devices are introduced, the static and dynamic voltage equalization techniques for series connected devices are discussed, and the performance of these devices is compared.

18   Chapter 2   High-Power Semiconductor Devices

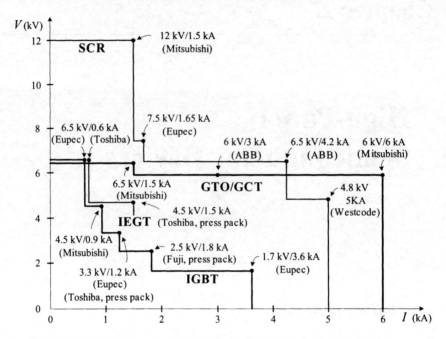

Figure 2.1-1 Voltage and current ratings of high-power semiconductor devices.

2.2 HIGH-POWER SWITCHING DEVICES

2.2.1 Diodes

High-power diodes can be generally classified into two types: (a) the general-purpose type for use in uncontrolled line-frequency rectifiers and (b) the fast recovery type used in voltage source converters as a freewheeling diode. These diodes are commercially available with two packaging techniques: press-pack and module diodes as shown in Fig. 2.2-1.

The device–heatsink assemblies for press-pack and module diodes are shown in Fig. 2.2-2. The press-pack diode features double-sided cooling with low thermal stress. For medium-voltage applications where a number of diodes may be connected in series, the diodes and their heatsinks can be assembled with just two bolts, leading to high power density and low assembly costs. This is one of the reasons for the continued popularity of press-pack semiconductors in the medium-voltage drives. The modular diode has an insulated baseplate with single-sided cooling, where a number of diodes can be mounted onto a single piece of heatsink.

2.2.2 Silicon-Controlled Rectifier (SCR)

The SCR is a thyristor-based device with three terminals: gate, anode, and cathode. It can be turned on by applying a pulse of positive gate current with a short duration

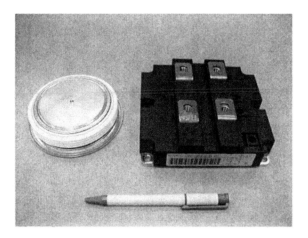

Figure 2.2-1 4.5-kV/0.8-kA press-pack and 1.7-kV/1.2-kA module diodes.

provided that it is forward-biased. Once the SCR is turned on, it is latched on. The device can be turned off by applying a negative anode current produced by its power circuit.

The SCR device can be used in phase-controlled rectifiers for PWM current source inverter-fed drives or load-commutated inverters for synchronous motor drives. Prior to the advent of self-extinguishable devices such as GTO and IGBT, the SCR was also used in forced commutated voltage source inverters.

The majority of high-power SCRs are of press-pack type as shown in Fig. 2.2-3. The SCR modules with an insulated baseplate are more popular for low- and medium-power applications.

Figure 2.2-4 shows the switching characteristics of the SCR device and typical waveforms for gate current i_G, anode current i_T, and anode–cathode voltage v_T.

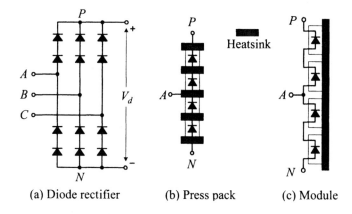

Figure 2.2-2 Device–heatsink assemblies for press-pack and module diodes.

20 Chapter 2 High-Power Semiconductor Devices

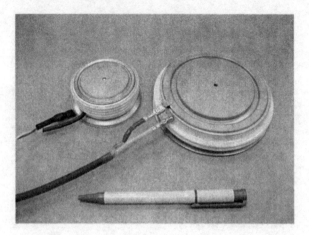

Figure 2.2-3 4.5-kV/0.8-kA and 4.5-kV/1.5-kA SCRs.

The turn-on process is initiated by applying a positive gate current i_G to the SCR gate. The turn-on behavior is defined by delay time t_d, rise time t_r and turn-on time t_{gt}.

The turn-off process is initiated by applying a negative current to the switch at time instant t_1, at which the anode current i_T starts to fall. The negative current is produced by the utility voltage when the SCR is used in a rectifier or by the load voltage in a load commutated inverter. The turn-off transient is characterized by re-

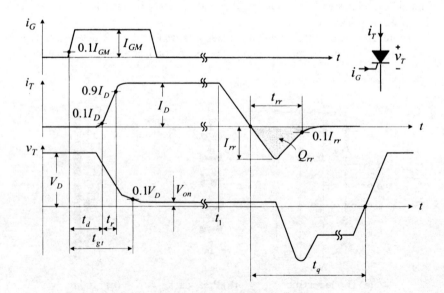

Figure 2.2-4 SCR switching characteristics.

verse recovery time t_{rr}, peak reverse recovery current I_{rr}, reverse recovery charge Q_{rr}, and turn-off time t_q.

Table 2.2-1 lists the main specifications of a 12-kV/1.5-kA SCR device, where V_{DRM} is the maximum repetitive peak off-state voltage, V_{RRM} is the maximum repetitive peak reverse voltage, I_{TAVM} is the maximum average on-state current, and I_{TRMS} is the maximum rms on-state current. The turn-on time t_{gt} is 14 μs and the turn-off time t_q is 1200 μs. The rates of anode current rise di_T/dt at turn-on and device voltage rise dv_T/dt at turn-off are important parameters for converter design. To ensure a proper and reliable operation, the maximum limits for the di_T/dt and dv_T/dt must not be exceeded. The reverse recovery charge Q_{rr} is normally a function of reverse recovery time t_{rr} and reverse recovery current I_{rr}. To reduce the power loss at turn-off, the SCR with a low value of Q_{rr} is preferred.

2.2.3 Gate Turn-Off (GTO) Thyristor

The gate turn-off (GTO) thyristor is a self-extinguishable device that can be turned off by a negative gate current. The GTOs are normally of press-pack design as shown in Fig. 2.2-5, and the modular design is not commercially available. Several manufacturers offer GTOs up to a rated voltage of 6 kV with a rated current of 6 kA.

The GTO can be fabricated with symmetrical or asymmetrical structures. The symmetric GTO has reverse voltage-blocking capability, making it suitable for current source converters. Its maximum repetitive peak off-state voltage V_{DRM} is approximately equal to its maximum repetitive peak reverse voltage V_{RRM}. The asymmetric GTO is generally used in voltage source converters where the reverse voltage-blocking capability is not required. The value of V_{RRM} is typically around 20 V, much lower than V_{DRM}.

The switching characteristics of the GTO thyristor are shown in Fig. 2.2-6, where i_T and v_T are the anode current and anode–cathode voltage, respectively. The GTO turn-on behavior is measured by delay time t_d and rise time t_r. The turn-off transient is characterized by storage time t_s, fall time t_f, and tail time t_{tail}. Some manufacturers provide only turn-on time t_{gt} ($t_{gt} = t_d + t_r$) and turn-off time t_{gq} ($t_{gq} = t_s + t_f$) in their datasheets. The GTO is turned on by a pulse of positive gate current of a few hundred milliamps. Its turn-off process is initiated by a negative gate cur-

Table 2.2-1 Main Specifications of a 12-kV/1.5-kA SCR

Maximum Ratings	V_{DRM}	V_{RRM}	I_{TAVM}	I_{TRMS}	—
	12,000 V	12,000 V	1500 A	2360 A	—
Switching Characteristics	Turn-on Time	Turn-off Time	di_T/dt	dv_T/dt	Q_{rr}
	t_{gt} = 14 μs	t_q = 1200 μs	100 A/μs	2000 V/μs	7000 μC
Part number: FT1500AU-240 (Mitsubishi)					

Figure 2.2-5 4.5-kV/0.8-kA and 4.5-kV/1.5-kA GTO devices.

rent. To ensure a reliable turn-off, the rate of change of the negative gate current di_{G2}/dt must meet with the specification set by the device manufacturer.

Table 2.2-2 gives the main specifications of a 4500-V/4000-A asymmetrical GTO device, where V_{DRM}, V_{RRM}, I_{TAVM}, and I_{TRMS} have the same definitions as those for the SCR device. It is worth noting that the current rating of a 4000-A GTO is defined by I_{TGQM}, which is the maximum repetitive controllable on-state current, not by the average current I_{TAVM}. The turn-on delay time t_d and rise time t_r are 2.5 μs and 5.0 μs while the storage time t_s and fall time t_f at turn-off are 25 μs and 3 μs,

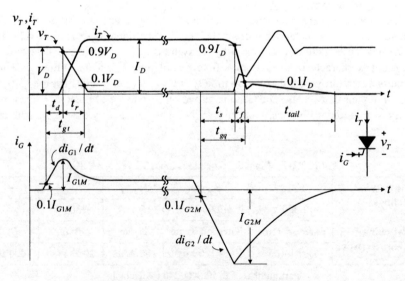

Figure 2.2-6 GTO switching characteristics.

2.2 High-Power Switching Devices

Table 2.2-2 Main Specifications of a 4.5-kV/4-kA Asymmetrical GTO

Maximum Rating	V_{DRM}	V_{RRM}	I_{TGQM}	I_{TAVM}	I_{TRMS}	—
	4500 V	17 V	4000 A	1000 A	1570 A	—
Switching Characteristics	Turn-on Switching	Turn-off Switching	di_T/dt	dv_T/dt	di_{G1}/dt	di_{G2}/dt
	$t_d = 2.5\ \mu s$	$t_s = 25.0\ \mu s$	500 A/μs	1000 V/μs	40 A/μs	40 A/μs
	$t_r = 5.0\ \mu s$	$t_f = 3.0\ \mu s$				
On-State Voltage	$V_{T(on\text{-}state)} = 4.4$ V at $I_T = 4000$ A					
Part number: 5SGA 40L4501 (ABB)						

respectively. The maximum rates of rise of the anode current, gate current, and device voltage are also given in the table.

The GTO thyristor features high on-state current density and high blocking voltage. However, the GTO device has a number of drawbacks, including (a) bulky and expensive turn-off snubber circuits due to low dv_T/dt, (b) high switching and snubber losses, and (c) complex gate driver. It also needs a turn-on snubber to limit di_T/dt.

2.2.4 Gate-Commutated Thyristor (GCT)

The gate-commutated thyristor (GCT), also known as integrated gate-commutated thyristor (IGCT), is developed from the GTO structure [2, 3]. Over the past few years, the industry has seen the GTO thyristor being replaced by the GCT device. The GCT has become the device of choice for medium voltage drives due to its features such as snubberless operation and low switching loss.

The key GCT technologies include significant improvements in silicon wafer, gate driver, and device packaging. The GCT wafer is much thinner than the GTO wafer, leading to a reduction in on-state power loss. As shown in Fig. 2.2-7, a special gate driver with ring-gate packaging provides an extremely low gate inductance (typically < 5 nH) that allows the GCT to operate without snubber circuits. The rate of gate current change at turn-off is normally greater than 3000 A/μs instead of around 40 A/μs for the GTO device. Since the gate driver is an integral part of the GCT, the user only needs to provide the gate driver with a 20- to 30-V dc power supply and connect the driver to the system controller through two fiber-optic cables for on/off control and device fault diagnostics.

Several manufacturers offer GCT devices with ratings up to 6 kV/6 kA. 10-kV GCTs are technically possible, and the development of this technology depends on the market needs [4].

The GCT devices can be classified into asymmetrical, reverse-conducting and symmetrical types as shown in Table 2.2-3. The asymmetric GCT is generally used

Figure 2.2-7 6.5-kV/1.5-kA symmetrical GCT.

in voltage source converters where the reverse voltage-blocking capability is not required. The reverse-conducting GCT integrates the freewheeling diode into one package, resulting in a reduced assembly cost. The symmetric GCT is normally for use in current source converters.

Figure 2.2-8 shows the typical switching characteristics of the GCT device, where the delay time t_d, rise time t_r, storage time t_s and fall time t_f are defined in the same way as those for the GTO. Note that some semiconductor manufacturers may define the switching times differently or use different symbols. The waveform for the gate current i_G is given as well, where the rate of gate current change di_{G2}/dt at turn-off is substantially higher than that for the GTO.

Table 2.2-4 gives the main specifications of a 6000-V/6000-A asymmetrical GCT, where the maximum repetitive controllable on-state current I_{TQRM} is 6000 A. The turn-on and turn-off times are much faster than those for the GTO. In particular, the storage time t_s is only 3 μs in comparison with 25 μs for the 4000-A GTO device in Table 2.2-2. The maximum dv_T/dt can be as high as 3000 V/μs.

Table 2.2-3 GCT Device Classification

Type	Antiparallel Diode	Blocking Voltage	Example (6000V GCT)	Applications
Asymmetrical GCT	Excluded	$V_{RRM} \ll V_{DRM}$	V_{DRM} = 6000 V V_{RRM} = 22 V	For use in voltage source converters with antiparallel diodes.
Reverse-conducting GCT	Included	$V_{RRM} \approx 0$	V_{DRM} = 6000 V	For use in voltage source converters.
Symmetrical GCT (reverse blocking)	Not required	$V_{RRM} \approx V_{DRM}$	V_{DRM} = 6000 V V_{RRM} = 6500 V	For use in current source converters.

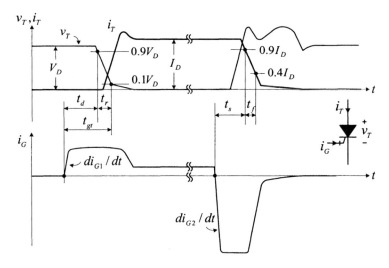

Figure 2.2-8 GCT switching characteristics.

The maximum rate of gate current change, di_{G2}/dt, can be as high as 10,000 A/μs, which helps to reduce the switching time at turn-off. The on-state voltage at $I_T = 6000$ A is only 4 V in comparison with 4.4 V at $I_T = 4000$ A for the GTO device.

The GCT device normally requires a turn-on snubber since the di_T/dt capability of the device is only around 1000 A/μs. Figure 2.2-9a shows a typical turn-on snubber circuit for voltage source converters [5]. The snubber inductor L_s limits the rate of anode current rise at the moment when one of the six GCTs is gated on. The energy trapped in the inductor is partially dissipated on the snubber resistor R_s. All six GCTs in the converter can share one snubber circuit. In current source converters, the snubber circuit takes a different form as shown in Fig. 2.2-9b, where a di/dt lim-

Table 2.2-4 Main Specifications of a 6KV/6KA Asymmetrical GCT

Maximum Rating	V_{DRM}	V_{RRM}	I_{TQRM}	I_{TAVM}	I_{TRMS}	—
	6000 V	22 V	6000 A	2000 A	3100 A	—
Switching Characteristics	Turn-on Switching	Turn-off Switching	di_T/dt	dv_T/dt	di_{G1}/dt	di_{G2}/dt
	$t_d < 1.0$ μs $t_r < 2.0$ μs	$t_s < 3.0$ μs t_f – N/A	1000 A/μs	3000 V/μs	200 A/μs	10,000 A/μs
On-state Voltage	$V_{T(on\text{-}state)} < 4$ V at $I_T = 6000$ A					
Part number: FGC6000AX120DS (Mitsubishi)						

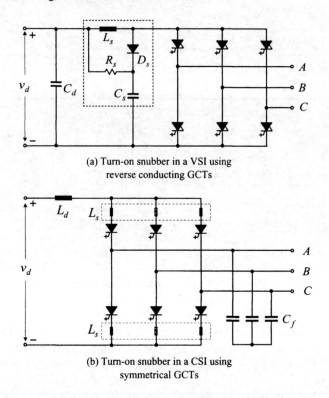

(a) Turn-on snubber in a VSI using reverse conducting GCTs

(b) Turn-on snubber in a CSI using symmetrical GCTs

Figure 2.2-9 Turn-on di/dt snubber for GCTs.

iting inductor of a few microhenries is required in each of the converter legs, but no other passive components are needed.

2.2.5 Insulated Gate Bipolar Transistor (IGBT)

The insulated gate bipolar transistor (IGBT) is a voltage-controlled device. It can be switched on with a +15 V gate voltage and turned off when the gate voltage is zero. In practice, a negative gate voltage of a few volts is applied during the device off period to increase its noise immunity. The IGBT does not require any gate current when it is fully turned on or off. However, it does need a peak gate current of a few amperes during switching transients due to the gate-emitter capacitance.

The majority of high-power IGBTs are of modular design as shown in Fig. 2.2-10. Press-pack IGBTs are also available on the market for assembly cost reduction and efficient cooling, but the selection of such devices is limited.

The typical switching characteristics of the IGBT device are shown in Fig. 2.2-11, where the turn-on delay time t_{don}, rise time t_r, turn-off delay time t_{doff}, and fall time t_f are defined. The waveforms for gate driver output voltage v_G, gate–emitter voltage v_{GE}, and collector current i_C are also given. The voltage v_{GE} is equal to v_G

2.2 High-Power Switching Devices

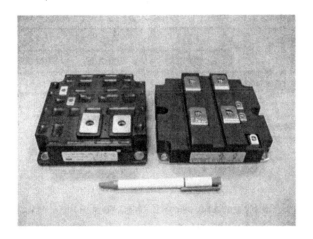

Figure 2.2-10 1.7-kV/1.2-kA and 3.3-kV/1.2-kA IGBT modules.

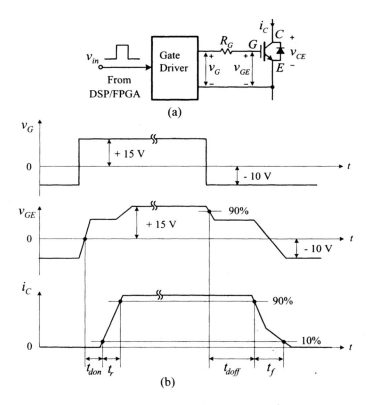

Figure 2.2-11 IGBT switching characteristics.

Table 2.2-5 Main Specifications of a 3.3-kV/1.2-kA IGBT

Maximum Rating	V_{CE}	I_C	I_{CM}	—
	3300 V	1200 A	2400 A	—
Switching Characteristics	t_{don}	t_r	t_{doff}	t_f
	0.35 μs	0.27 μs	1.7 μs	0.2 μs
Saturation Voltage	$I_{CE\,sat}$ = 4.3 V at I_C = 1200 A			
Part number: FZ1200 R33 KF2 (Eupec)				

after the IGBT is fully turned on or off. These two voltages, however, are not the same during switching transients due to the gate-emitter capacitance. The gate resistor R_G is normally required to adjust the device switching speed and to limit the transient gate current.

Table 2.2-5 gives the main specifications of a 3.3-kV/1.2-kA IGBT, where V_{CE} is the rated collector–emitter voltage, I_C is the rated dc collector current and I_{CM} is the maximum repetitive peak collector current. The IGBT has superior switching characteristics. It can be turned on within 1 μs and turned off within 2 μs.

The IGBT device features simple gate driver, snubberless operation, high switching speed, and modular design with insulated baseplate. More importantly, the IGBT can operate in the active region. Its collector current can be controlled by the gate voltage, providing an effective means for reliable short-circuit protection and active control of dv/dt and overvoltage at turn-off.

The construction of a medium-voltage converter with series connected IGBT modules should consider a number of issues such as efficient cooling arrangements, optimal dc bus-bar design, and stray capacitance of baseplates to ground. In contrast, press-pack IGBTs allow direct series connection, where the mounting and cooling techniques developed for press-pack thyristors can be utilized.

2.2.6 Other Switching Devices

There are a number of other semiconductor devices, including power MOSFET, emitter turn-off thyristor (ETO) [6], MOS-controlled thyristor (MCT), and static induction thyristor (SIT). However, they have not gained significant importance in high-power applications. The injection enhanced gate transistor (IEGT) seems to be a promising new switching device for high-power converters [7].

2.3 OPERATION OF SERIES-CONNECTED DEVICES

In medium voltage drives, switching devices are normally connected in series. It is not necessary to parallel the devices since the current capacity of a single device is usually sufficient. For instance, in a 6.6-kV 10-MW drive the rated motor current is only 880A in comparison with the current rating of a 6000A GCT or 3600A IGBT.

Since the series-connected devices and their gate drivers may not have exactly the same static and dynamic characteristics, they may not equally share the total voltage in the blocking mode or during switching transients. The main task for the series-connected switches is to ensure equal voltage-sharing under both static and dynamic conditions.

2.3.1 Main Causes of Voltage Unbalance

The static voltage unbalance is mainly caused by the difference in the off-state leakage current I_{lk} of series-connected switches. Furthermore, the leakage current is a function of device junction temperature and operating voltage. The causes of the dynamic voltage unbalance can be divided into two groups: (a) unbalance due to the difference in device switching behavior and (b) unbalance caused by the difference in gate signal delays between the system controller and the switches. Table 2.3-1 summarizes the main causes of unequal voltage distribution, where Δ represents the discrepancies between series-connected devices.

2.3.2 Voltage Equalization for GCTs

(1) Static Voltage Equalization. Figure 2.3-1a shows a commonly used method for static voltage equalization, where each switch is protected by a parallel resistor R_p. Its resistance can be determined by an empirical equation

$$R_p = \frac{\Delta V_T}{\Delta I_{lk}} \qquad (2.3\text{-}1)$$

Table 2.3-1 Main Causes of Unequal Voltage Distribution Between Series-Connected Devices

Type	Causes of Voltage Unbalance		
Static voltage unbalance	ΔI_{lk}: Device off-state leakage current		
	ΔT_j: Junction temperature		
Dynamic voltage unbalance	Device	Δt_{don}:	Turn-on delay time
		Δt_{doff}:	Turn-off delay time (IGBT)
		Δt_s:	Storage time (GCT)
		ΔQ_{rr}:	Reverse recovery charge
		ΔT_j:	Junction temperature
	Gate driver	Δt_{GDon}:	Gate driver turn-on delay time
		Δt_{GDoff}:	Gate driver turn-off delay time
		ΔL_{wire}:	Wiring inductance between the gate driver output and the device gate

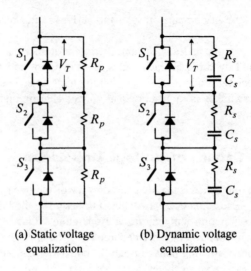

Figure 2.3-1 Passive voltage equalization techniques.

where ΔV_T is the desired maximum voltage discrepancy between the series switches and ΔI_{lk} is the allowable tolerance for the off-state leakage current. Equation (2.3-1) is valid for both asymmetrical and symmetrical GCTs, and the value of R_p is normally between 20 kΩ and 100 kΩ [8].

(2) Dynamic Voltage Equalization. For the dynamic voltage equalization, three modes of GCT operation need to be considered:

- Turn-on transient
- Turn-off transient by gate commutation
- Turn-off transient by natural commutation (for symmetrical GCT only)

The first two operating modes are for asymmetrical and reverse-conducting GCTs used in voltage source converters, while all three operating modes are applicable to symmetrical GCTs in current source converters.

To ensure equal dynamic voltage sharing, the following techniques can be employed:

- Use devices of one production lot to minimize Δt_{don}, Δt_s and ΔQ_{rr}.
- Match the device switching characteristics to minimize Δt_{don}, Δt_s and ΔQ_{rr}.
- Make the device cooling condition identical to minimize ΔT_j.
- Design symmetrical gate drivers to minimize Δt_{GDon} and Δt_{GDoff}.
- Place the gate drivers symmetrically to minimize ΔL_{wire}.

The implementation of the above-mentioned techniques can help to reduce the device voltage unbalance during switching transients, but does not guarantee a satisfactory result. The series connected switches are often protected by RC snubber circuits shown in Fig. 2.3-1b.

The snubber capacitor C_s should be sized to minimize the effect of the delay time inconsistency on the GCT voltage equalization. Since the turn-on delay time t_{don} is normally much shorter than the storage time t_s at turn-off, the requirements for turn-off dominate. The value of C_s can be found from an empirical equation

$$C_s = \frac{\Delta t_{delay} \times I_{T\max}}{\Delta V_{T\max}} \qquad (2.3\text{-}2)$$

where Δt_{delay} is the maximum tolerance in the total turn-off delay time including t_s and the delay time caused by gate drivers, $I_{T\max}$ is the maximum anode current to be commutated, and $\Delta V_{T\max}$ is the maximum allowed voltage deviation between the series switches [8].

The GCTs used in the current source converter may be turned off by natural commutation, where the device is commutated by a negative anode current produced by its power circuit. The commutation process is similar to that of an SCR device. The dominant factor affecting the dynamic voltage unbalance in this case is the discrepancy in the GCT reverse recovery charge ΔQ_{rr}. This adds another criterion for choosing the capacitance value:

$$C_s = \frac{\Delta Q_{rr}}{\Delta V_{T\max}} \qquad (2.3\text{-}3)$$

The value of C_s is normally in the range of 0.1 to 1 μF for the GCT devices, much lower than that for the GTOs. The snubber resistance R_s should (a) be sized such that it should be small enough to allow fast charging and discharging of the snubber capacitor to accommodate the short pulsewidths of the PWM operation and (b) be large enough to limit the discharging current that flows through the GCT at turn-on. A good compromise should be made.

2.3.3 Voltage Equalization for IGBTs

The static and dynamic voltage equalization techniques for the GCT devices can be equally applied to the IGBTs. In addition, an active overvoltage clamping scheme can be implemented to limit the collector–emitter voltage during switching transients. This scheme is invalid for the GCTs due to the latching mechanism of the thyristor structure.

Figure 2.3-2 illustrates the principle of an active overvoltage clamping scheme [9]. The collector–emitter voltage v_{CE} of each IGBT is detected and compared with a reference voltage $V_{\max}^*$ that is the maximum allowed voltage for the device. The difference Δv is sent to a comparator. If the detected v_{CE} is lower than $V_{\max}^*$ at turn-

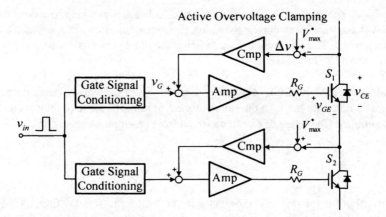

Figure 2.3-2 Principle of active overvoltage clamping for series-connected IGBTs.

off, the output of the comparator is zero and the operation of the device is not affected. At the moment that v_{CE} tends to exceed V^*_{max}, $|\Delta v|$ is added to the gate signal v_G, forcing v_{CE} to decrease. Through the feedback control in the IGBT active region, v_{CE} will be clamped to the value set by V^*_{max} during switching transients, effectively protecting the device from overvoltage. However, this is achieved at the expense of an increase in the device switching loss.

2.4 SUMMARY

The chapter focuses on commonly used high-power semiconductor devices including SCRs, GTOs, GCTs, and IGBTs. Their switching characteristics are introduced and main specifications are discussed. Since these devices are often connected in series for high-power medium-voltage applications, the static and dynamic voltage equalization techniques are elaborated. To summarize, a qualitative comparison for the GTO, GCT, and IGBT devices is given as follows.

Item	GTO	GCT	IGBT
Maximum voltage and current ratings	High	High	Low
Packaging	Press pack	Press pack	Module or press pack
Switching speed	Slow	Moderate	Fast
Turn-on (di/dt) snubber	Required	Required	Not required
Turn-off (dv/dt) snubber	Required	Not required	Not required
Active overvoltage clamping	No	No	Yes
Active di/dt and dv/dt control	No	No	Yes
Active short-circuit protection	No	No	Yes

Item	GTO	GCT	IGBT
On-state loss	Low	Low	High
Switching loss	High	Medium	Low
Behavior after destruction	Short-circuited	Short-circuited	Open-circuited
Gate driver	Complex, separate	Complex, integrated	Simple, compact
Gate driver power consumption	High	Medium	Low

REFERENCES

1. S. Bernet, Recent Developments of High Power Converters for Industry and Traction Applications, *IEEE Transactions on Power Electronics,* Vol. 15, No. 6, pp. 1102–1117, 2000.
2. P. K. Steimer, H. E. Gruning, et al., IGCT—A New Emerging Technology for High Power Low Cost Inverters, *IEEE Industry Application Magazine,* pp. 12–18, July/August, 1999.
3. H. M. Stillman, IGCTs—Megawatt Power Switches for Medium Voltage Applications, *ABB Review,* No. 3, pp. 12–17, 1997.
4. S. Eicher, S. Bernet, et al., The 10 kV IGCT—A New Device for Medium Voltage Drives, *IEEE Industry Applications Conference,* pp. 2859–2865, 2000.
5. A. Nagel, S. Bernet, et al., Characterization of IGCTs for Series Connected Operation, *IEEE Industry Applications Conference (IAS),* Vol. 3, pp. 1923–1929, 2000.
6. K. Motto, Y. Li, et al., High Frequency Operation of a Megawatt Voltage Source Inverter Equipped with ETOs, *IEEE Applied Power Electronics Conference (PESC),* Vol. 2, pp. 924–930, 2001.
7. T. Ogura, H. Ninomiya, et al., 4.5-kV Injection-Enhanced Gate Transistors (IEGTs) with High Turn-Off Ruggedness, *IEEE Transactions on Electron Devices,* Vol. 51, No. 4, pp. 636–641, 2004.
8. N. R. Zargari, S. C. Rizzo, et at., A New Current-Source Converter Using a Symmetric Gate-Commutated Thyristor (SGCT), *IEEE Transactions on Industry Applications,* Vol. 37, No. 3, pp. 896–903, 2001.
9. M. Bruckmann, R. Sommer, et al., Series Connection of High Voltage IGBT Modules, *IEEE Industry Applications Society (IAS) Conference,* pp. 1067–1072, 1998.

Part Two

Multipulse Diode and SCR Rectifiers

Chapter 3

Multipulse Diode Rectifiers

3.1 INTRODUCTION

In an effort to comply with the stringent harmonic requirements set by North American and European standards such as IEEE Standard 519-1992, major high-power drive manufacturers around the world are increasingly using multipulse diode rectifiers in their drives as front-end converters [1–5]. The rectifiers can be configured as 12-, 18- and 24-pulse rectifiers, powered by a phase shifting transformer with a number of secondary windings. Each secondary winding feeds a six-pulse diode rectifier. The dc output of the six-pulse rectifiers is connected to a voltage source inverter.

The main feature of the multipulse rectifier lies in its ability to reduce the line current harmonic distortion. This is achieved by the phase shifting transformer, through which some of the low-order harmonic currents generated by the six-pulse rectifiers are canceled. In general, the higher the number of rectifier pulses, the lower the line current distortion is. The rectifiers with more than 30 pulses are seldom used in practice mainly due to increased transformer costs and limited performance improvements.

The multipulse rectifier has a number of other features. It normally does not require any LC filters or power factor compensators, which leads to the elimination of possible LC resonances. The use of the phase-shifting transformer provides an effective means to block common-mode voltages generated by the rectifier and inverter in medium voltage drives, which would otherwise appear on motor terminals, leading to a premature failure of winding insulation [6,7].

The multipulse diode rectifiers can be classified into two types:

- **Series-type multipulse rectifiers,** where all the six-pulse rectifiers are connected in series on their dc side. In the medium-voltage drives, the series-type rectifier can be used as a front end for the inverter that requires a single dc supply such as three-level neutral point clamped (NPC) inverter and multilevel flying-capacitor inverter [1, 2].

- **Separate-type multipulse rectifiers,** where each of the six-pulse rectifiers feeds a separate dc load. This type of rectifier is suitable for use in a multilevel cascaded H-bridge inverter that requires a multiple units of isolated dc supplies [4, 5].

This chapter starts with an introduction to the six-pulse diode rectifier, followed by a detailed analysis on the series- and separate-type multipulse rectifiers. The rectifier input power factor and line current THD are investigated, and results are summarized in a graphical format. The phase-shifting transformer required by the multipulse rectifiers will be dealt with in Chapter 5.

3.2 SIX-PULSE DIODE RECTIFIER

3.2.1 Introduction

The circuit diagram of a simplified six-pulse diode rectifier is shown in Fig. 3.2-1, where v_a, v_b, and v_c are the phase voltages of the utility supply. For medium-voltage applications, each diode in the rectifier may be replaced by two or more diodes in series. To simplify the analysis, all the diodes are assumed to be ideal (no power losses or on-state voltage drop).

Figure 3.2-2 shows a set of voltage and current waveforms of the rectifier. The phase voltages of the utility supply are defined by

$$v_a = \sqrt{2}\, V_{PH} \sin(\omega t)$$

$$v_b = \sqrt{2}\, V_{PH} \sin(\omega t - 2\pi/3) \qquad (3.2\text{-}1)$$

$$v_c = \sqrt{2}\, V_{PH} \sin(\omega t - 4\pi/3)$$

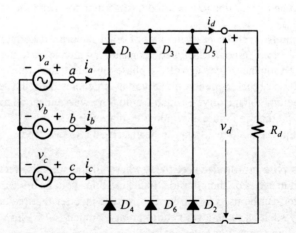

Figure 3.2-1 Six-pulse diode rectifier with a resistive load.

where V_{PH} is the rms value of the phase voltage and ω is the angular frequency of the utility supply, given by $\omega = 2\pi f$. The line-to-line voltage v_{ab} can be calculated by

$$v_{ab} = v_a - v_b = \sqrt{2}\, V_{LL} \sin(\omega t + \pi/6) \qquad (3.2\text{-}2)$$

where V_{LL} is the rms value of the line-to-line voltage relating to the phase voltage by $V_{LL} = \sqrt{3}\, V_{PH}$.

The waveform of the line current i_a has two humps per half-cycle of the supply frequency. During interval *I*, v_{ab} is higher than the other two line-to-line voltages. Diodes D_1 and D_6 are forward-biased and thus turned on. The dc voltage v_d is equal to v_{ab}, and the line current i_a equals v_{ab}/R_d. During interval *II*, D_1 and D_2 conduct, and $i_a = v_{ac}/R_d$. Similarly, the waveform of i_a in the negative half-cycle can be obtained. The other two line currents, i_b and i_c, have the same waveform as i_a, but lag i_a by $2\pi/3$ and $4\pi/3$, respectively.

The dc voltage v_d contains six pulses (humps) per cycle of the supply frequency. The rectifier is, therefore, commonly known as a six-pulse rectifier. The average value of the dc voltage can be calculated by

$$V_{do} = \frac{\text{area } \mathbf{A}_1}{\pi/3} = \frac{1}{\pi/3}\int_{\pi/6}^{\pi/2}\sqrt{2}\, V_{LL}\sin(\omega t + \pi/6)d(\omega t) = \frac{3\sqrt{2}}{\pi} V_{LL} \approx 1.35 V_{LL} \qquad (3.2\text{-}3)$$

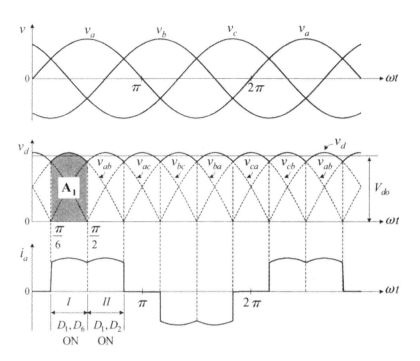

Figure 3.2-2 Waveforms of the six-pulse diode rectifier with a resistive load.

3.2.2 Capacitive Load

Figure 3.2-3a shows the circuit diagram of the six-pulse rectifier with a capacitive load, where L_s represents the total line inductance between the utility supply and the rectifier, including the equivalent inductance of the supply, the leakage inductances of isolation transformer if any, and the inductance of a three-phase reactor that is often added to the system in practice for the reduction of the line current THD. The dc filter capacitor C_d is assumed to be sufficiently large such that the dc voltage is ripple-free. Under this assumption, the capacitor and dc load can be replaced by a dc voltage source V_d as shown in Fig. 3.2-3b. The value of V_d varies slightly with the loading conditions. When the rectifier is lightly loaded, V_d is close to the peak of the ac line-to-line voltage and the dc current i_d may be discontinuous. With the increase in dc current, the voltage across the line inductance L_s increases, causing a reduction in V_d. When the dc current increases to a certain level, it becomes continuous and thus the rectifier operates in a continuous current mode.

(1) Discontinuous Current Operation. Fig. 3.2-4 illustrates the voltage and current waveforms of the rectifier operating under the light load conditions. Each of the three-phase line currents, i_a, i_b, and i_c, contains two pulses per half-cycle of the supply frequency. The rectifier operates in a discontinuous current mode since the dc current i_d falls to zero six times per cycle of the supply frequency.

For the convenience of discussion, Fig. 3.2-5 shows the expanded voltage and current waveforms of the rectifier. During interval $\theta_1 \leq \omega t < \theta_2$, the line-to-line voltage v_{ab} is higher than the dc voltage V_d, which turns D_1 and D_6 on. The dc current i_d increases from zero, and energy is stored in L_s. At θ_2, v_{ab} is equal to V_d, the

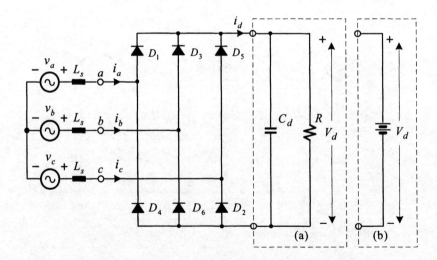

Figure 3.2-3 Six-pulse diode rectifier with a capacitive load.

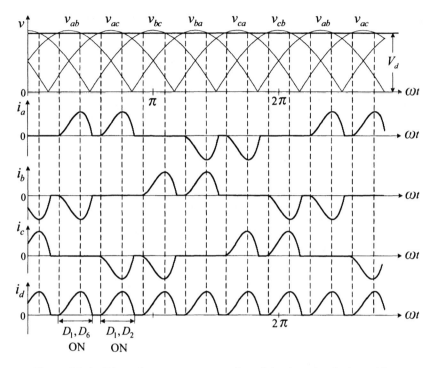

Figure 3.2-4 Discontinuous current operation of the six-pulse diode rectifier.

voltage across the line inductance L_s becomes zero, and i_d reaches its peak value I_p. When $\omega t > \theta_2$, v_{ab} is lower than V_d, and L_s starts to release its stored energy to the load through D_1 and D_6. Both diodes remain on until $\omega t = \theta_3$, at which i_d falls to zero and the energy stored in L_s is completely discharged. During interval $\theta_4 \leq \omega t < \theta_5$, v_{ac} is higher than V_d, and D_1 is turned on again together with D_2. Obviously,

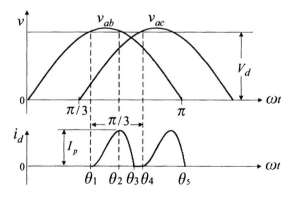

Figure 3.2-5 Details of the dc current waveform in the six-pulse diode rectifier.

each diode conducts twice per cycle of the supply frequency. The diode conduction angle is given by

$$\theta_c = 2(\theta_3 - \theta_1), \qquad 0 \le \theta_c \le 2\pi/3 \tag{3.2-4}$$

At θ_1 and θ_2, the line-to-line voltage v_{ab} is equal to V_d, from which

$$\theta_1 = \sin^{-1}\left(\frac{V_d}{\sqrt{2}\,V_{LL}}\right) \tag{3.2-5}$$

and

$$\theta_2 = \pi - \theta_1 \tag{3.2-6}$$

The total voltage across the two line inductances in phases a and b when D_1 and D_6 are on can be expressed as

$$2L_s \frac{di_d}{dt} = v_{ab} - V_d \quad \text{for } \theta_1 \le \omega t < \theta_3 \tag{3.2-7}$$

from which

$$i_d(\theta) = \frac{1}{2\omega L_s} \int_{\theta_1}^{\theta} (\sqrt{2}\,V_{LL}\sin(\omega t) - V_d)\,d(\omega t)$$

$$= \frac{1}{2\omega L_s}(\sqrt{2}\,V_{LL}(\cos\theta_1 - \cos\theta) + V_d(\theta_1 - \theta)) \tag{3.2-8}$$

The peak dc current can be calculated by substituting θ_2 into (3.2-8)

$$\hat{I}_d = \frac{1}{2\omega L_s}(\sqrt{2}\,V_{LL}(\cos\theta_1 - \cos\theta_2) + V_d(\theta_1 - \theta_2)) \tag{3.2-9}$$

The average dc current can be calculated by

$$I_d = \frac{1}{\pi/3} \int_{\theta_1}^{\theta_3} i_d(\theta)\,d(\theta) \tag{3.2-10}$$

Substituting the condition of $i_d(\theta_3) = 0$ into (3.2-8) yields

$$\frac{V_d}{\sqrt{2}\,V_{LL}} = \frac{\cos\theta_3 - \cos\theta_1}{\theta_1 - \theta_3} \tag{3.2-11}$$

from which θ_3 can be calculated for a given V_{LL} and V_d. It is interesting to note that the angles θ_1, θ_2, and θ_3 are a function of V_{LL} and V_d only, irrelevant to the line inductance L_s.

(2) Continuous Current Operation. As discussed earlier, the dc voltage V_d of the rectifier decreases with the increase in dc load current. The decrease in V_d makes θ_3 and θ_4 in Fig. 3.2-5 move toward each other. When θ_3 and θ_4 start to overlap, the dc current i_d becomes continuous.

Figure 3.2-6 shows the waveforms of the rectifier operating with a continuous dc current. During interval *I*, a positive i_a keeps D_1 conducting while a negative i_c keeps D_2 on. The dc current is given by $i_d = i_a = -i_c$.

Interval *II* is the commutation period, during which the current flowing in D_1 is commutated to D_3. The commutation is initiated by a forward-biased voltage on D_3, which turns it on. Due to the presence of the line inductance L_s, the commutation process cannot complete instantly. It takes a finite moment for current i_b in D_3 to build up and current i_a in D_1 to fall. During this period, three diodes (D_1, D_2 and D_3) conduct simultaneously, and the dc current $i_d = i_a + i_b = -i_c$. The commutation period ends at the end of interval *II*, at which i_a decreases to zero and D_1 is turned off.

During interval *III*, diodes D_2 and D_3 are on, and the dc current is $i_d = i_b = -i_c$. The diode conduction angle θ_c is $2\pi/3 + \gamma$, where γ is the commutation interval. Compared with discontinuous current operation, the rectifier operating under a continuous current mode draws a line current with less harmonic distortion. The details of the line current distortion are discussed in the following sections.

3.2.3 Definition of THD and PF

Assume that the phase voltage v_a of the utility supply is sinusoidal

$$v_a = \sqrt{2} V_a \sin \omega t \qquad (3.2-12)$$

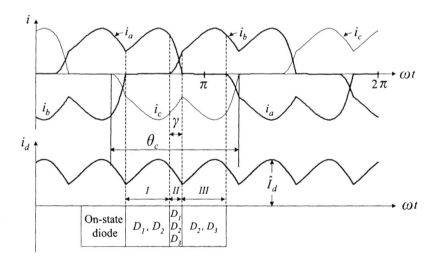

Figure 3.2-6 Current waveforms of the six-pulse diode rectifier operating in a continuous current mode.

The line current i_a drawn by a rectifier is generally periodical but nonsinusoidal. The line current can be expressed by a Fourier series

$$i_a = \sum_{n=1,2,3,...}^{\infty} \sqrt{2}\, I_{an}(\sin(\omega_n t) - \phi_n) \qquad (3.2\text{-}13)$$

where n is the harmonic order, I_{an} and ω_n are the rms value and angular frequency of the nth harmonic current, and ϕ_n is the phase displacement between V_a and I_{an}, respectively.

The rms value of the distorted line current i_a can be calculated by

$$I_a = \left(\frac{1}{2\pi} \int_0^{2\pi} (i_a)^2 d(\omega t) \right)^{1/2} = \left(\sum_{n=1,2,3,...}^{\infty} I_{an}^2 \right)^{1/2} \qquad (3.2\text{-}14)$$

The total harmonic distortion is defined by

$$\text{THD} = \frac{\sqrt{I_a^2 - I_{a1}^2}}{I_{a1}} \qquad (3.2\text{-}15)$$

where I_{a1} is the rms value of the fundamental current. The per-phase average power delivered from the supply to the rectifier is

$$P = \frac{1}{2\pi} \int_0^{2\pi} v_a \times i_a d(\omega t) \qquad (3.2\text{-}16)$$

Substituting (3.2-12) and (3.2-13) into (3.2-16) yields

$$P = V_a I_{a1} \cos \phi_1 \qquad (3.2\text{-}17)$$

where ϕ_1 is the phase displacement between V_a and I_{a1}. The per-phase apparent power is given by

$$S = V_a I_a \qquad (3.2\text{-}18)$$

The input power factor is defined as

$$\text{PF} = \frac{P}{S} = \frac{I_{a1}}{I_a} \cos \phi_1 = \text{DF} \times \text{DPF} \qquad (3.2\text{-}19)$$

where DF is the distortion factor and DPF is the displacement power factor, given by

$$\begin{aligned} \text{DF} &= I_{a1}/I_a \\ \text{DPF} &= \cos \phi_1 \end{aligned} \qquad (3.2\text{-}20)$$

For a given THD and DPF, the power factor can also be calculated by

$$\text{PF} = \frac{\text{DPF}}{\sqrt{1+\text{THD}^2}} \tag{3.2-21}$$

3.2.4 Per-Unit System

It is convenient to use per-unit system for the analysis of power converter systems. Assume that the converter system under investigation is three-phase balanced with a rated apparent power S_R and rated line-to-line voltage V_{LL}. The base voltage, which is the rated phase voltage of the system, is given by

$$V_B = \frac{V_{LL}}{\sqrt{3}} \tag{3.2-22}$$

The base current and impedance are then defined as

$$I_B = \frac{S_R}{3V_B} \quad \text{and} \quad Z_B = \frac{V_B}{I_B} \tag{3.2-23}$$

The base frequency is

$$\omega_B = 2\pi f \tag{3.2-24}$$

where f is the nominal frequency of the utility supply or the rated output frequency of an inverter. The base inductance and capacitance can be found from

$$L_B = \frac{Z_B}{\omega_B} \quad \text{and} \quad C_B = \frac{1}{\omega_B Z_B} \tag{3.2-25}$$

Consider a three-phase diode rectifier rated at 4160 V, 60 Hz, and 2 MVA. The base current I_B of the rectifier is 277.6 A and the base inductance L_B is 22.9 mH. Assuming that the rectifier has a line inductance of 2.29 mH per phase and draws a line current of 138.8 A, the corresponding per unit value for the inductance and current is 0.1 per unit (pu) and 0.5 pu, respectively.

3.2.5 THD and PF of Six-Pulse Diode Rectifier

Two typical waveforms of the line current drawn by the six-pulse diode rectifier are shown in Fig. 3.2-7. When the rectifier operates under the light load conditions with the fundamental line current $I_{a1} = 0.2$ pu, the line current waveform is somewhat spiky. The line current waveform contains two separate pulses per half-cycle of the supply frequency, which makes the dc current discontinuous. With the rectifier operating at the rated condition (I_{a1} = 1.0 pu), the two current pulses are partially overlapped, which leads to a continuous dc current.

The harmonic spectrum of the line current waveform is shown in Fig. 3.2-7c. The line current i_a does not contain any even-order harmonics since the current waveform is of **half-wave symmetry**, defined by $f(\omega t) = -f(\omega t + \pi)$. It does not contain any triplen (**zero sequence**) harmonic currents either due to a balanced three-phase system. The dominant harmonics, such as the 5th and 7th, usually have a much higher magnitude than other harmonics. The line current THD is a function of the fundamental current I_{a1}, which is 75.7% at I_{a1} = 0.2 pu and 32.7% at I_{a1} = 1.0 pu.

The THD and PF curves of the six-pulse diode rectifier are shown in Fig. 3.2-8, where the fundamental current I_{a1} varies from 0.1 pu to 1 pu and the line inductance L_s changes from 0.05 pu to 0.15 pu. With the increase of the line current, its THD decreases while the overall power factor (PF) of the rectifier increases. The increase in PF is mainly due to THD reduction, which improves the distortion power factor

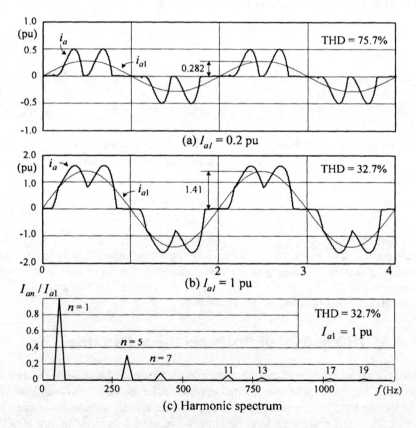

Figure 3.2-7 Line current waveform and harmonic content of the six-pulse diode rectifier (L_s = 0.05 pu).

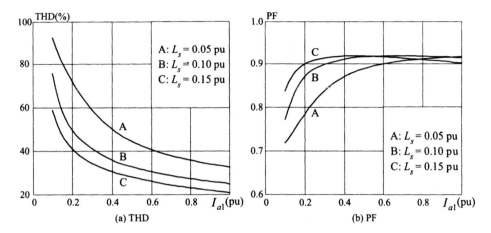

Figure 3.2-8 Calculated THD and PF of the six-pulse diode rectifier.

of the rectifier. For a given value of THD and PF, the displacement power factor DPF can be found from

$$\text{DPF} = \cos\phi_1 = \text{PF}\sqrt{1 + \text{THD}^2} \qquad (3.2\text{-}26)$$

As mentioned earlier, the low-order dominant harmonics in the line current are of high magnitude. An effective approach to reducing the line current THD is, therefore, to remove these dominant harmonics from the system. This can be achieved by using multipulse rectifiers.

3.3 SERIES-TYPE MULTIPULSE DIODE RECTIFIERS

In this section, the configuration of 12-, 18-, and 24-pulse series-type rectifiers is introduced. The THD and PF performance of these rectifiers are investigated through simulation and experiments.

3.3.1 12-Pulse Series-Type Diode Rectifier

(1) Rectifier Configuration. Figure 3.3-1 shows the typical configuration of a 12-pulse series-type diode rectifier. There are two identical six-pulse diode rectifiers powered by a phase-shifting transformer with two secondary windings. The dc outputs of the six-pulse rectifiers are connected in series. To eliminate low-order harmonics in the line current i_A, the line-to-line voltage v_{ab} of the wye-connected secondary winding is in phase with the primary voltage v_{AB} while the delta-connected secondary winding voltage $v_{\bar{a}\bar{b}}$ leads v_{AB} by

$$\delta = \angle v_{\bar{a}\bar{b}} - \angle v_{AB} = 30° \qquad (3.3\text{-}1)$$

The rms line-to-line voltage of each secondary winding is

$$V_{ab} = V_{\tilde{a}\tilde{b}} = V_{AB}/2 \qquad (3.3\text{-}2)$$

from which the turns ratio of the transformer can be determined by

$$\frac{N_1}{N_2} = 2 \quad \text{and} \quad \frac{N_1}{N_3} = \frac{2}{\sqrt{3}} \qquad (3.3\text{-}3)$$

The inductance L_s represents the total line inductance between the utility supply and the transformer, and L_{lk} is the total leakage inductance of the transformer referred to the secondary side. The dc filter capacitor C_d is assumed to be sufficiently large such that the dc voltage V_d is ripple-free.

Figure 3.3-1b shows the simplified diagram of the 12-pulse diode rectifier. The transformer winding is represented by sign "Y" or "Δ" enclosed by a circle, where 'Y' denotes a three-phase wye-connected winding while 'Δ' represents a delta-connected winding.

(2) Current Waveforms. Figure 3.3-2 shows a set of simulated current waveforms of the rectifier operating under the rated conditions. The line inductance L_s is assumed to be zero, and the total leakage inductance L_{lk} is 0.05 pu, which is a typical value for a phase-shifting transformer.

The dc current i_d is continuous, containing 12 pulses per cycle of the supply frequency. At any time instant (excluding commutation intervals), the dc current i_d flows through four diodes simultaneously, two in the top six-pulse rectifier and two in the bottom rectifier. The dc current ripple is relatively low due to the series connection of the two six-pulse rectifiers, where the leakage inductances of the secondary windings can be considered in series.

The waveform of the line current i_a in the wye-connected secondary winding looks like a trapezoidal wave with four humps on the top. The waveform of $i_{\tilde{a}}$ in the delta-connected winding is identical to i_a except for a 30° phase displacement and is therefore not shown in the figure.

The currents i'_a and $i'_{\tilde{a}}$ in Fig. 3.3-2b are the secondary line currents i_a and $i_{\tilde{a}}$ referred to the primary side. Since both primary and top secondary windings are connected in wye, the waveform of the referred current i'_a is identical to that of i_a except that its magnitude is halved due to the turns ratio of the two windings. When $i_{\tilde{a}}$ is referred to the primary side, the referred current $i'_{\tilde{a}}$ does not keep the same waveform as $i_{\tilde{a}}$. The changes in waveform are caused by the phase displacement of the harmonic currents when they are referred from the delta-connected secondary winding to the wye-connected primary winding. It is the phase displacement that makes certain harmonics, such as the 5th and 7th, in $i'_{\tilde{a}}$ out of phase with those in i'_a. As a result, these harmonic currents are canceled in the transformer primary winding and do not appear in the primary line current, given by

$$i_A = i'_a + i'_{\tilde{a}} \qquad (3.3\text{-}4)$$

3.3 Series-Type Multipulse Diode Rectifiers

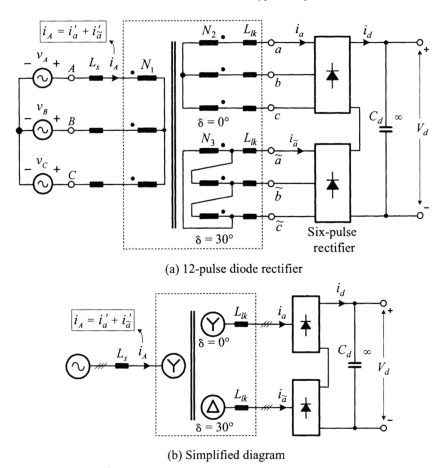

Figure 3.3-1 The 12-pulse series-type diode rectifier.

Figure 3.3-3 shows the harmonic spectrum of the rectifier currents in Fig. 3.3-2, where I'_{an}, $I'_{\tilde{a}n}$, and I_{An} are the nth order harmonic currents (rms) in i'_a, $i'_{\tilde{a}}$, and i_A, respectively. The harmonic content of the referred currents i'_a and $i'_{\tilde{a}}$ is identical, although their waveforms are quite different. This is understandable since the harmonic content should not alter when a secondary current is referred to the primary side. The magnitude of the 5th and 7th harmonics is 18.6% and 12.4%, respectively, which are much higher than other harmonics. The THD of the primary line current i_A is only 8.38% in comparison to 24.1% of the secondary line current i_a. The substantial reduction in THD is owing to the elimination of dominant harmonics by the phase-shifting transformer. The principle of harmonic cancellation by the phase-shifting transformer will be discussed in Chapter 5.

Figure 3.3-4 shows the waveforms measured from a 12-pulse diode rectifier operating under the rated conditions. The phase-shifting transformer has a total leak-

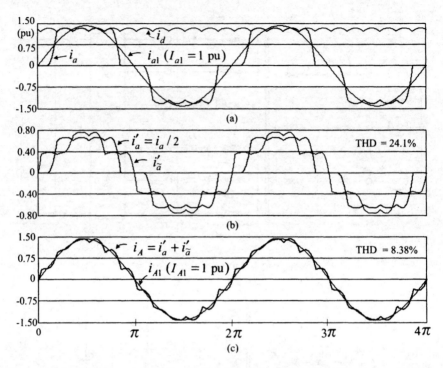

Figure 3.3-2 Current waveforms in the 12-pulse series-type rectifier ($L_s = 0$, $L_{lk} = 0.05$ pu, and $I_{A1} = 1.0$ pu).

age inductance of 0.045 pu and a voltage ratio of $V_{AB}/V_{ab} = V_{AB}/V_{\tilde{a}\tilde{b}} = 2.05$. The line inductance L_s between the utility supply and the transformer is negligible due to the high capacity of the supply and low power rating of the rectifier.

The measured secondary currents, i_a and $i_{\tilde{a}}$, are of a quasi-trapezoidal wave with a 30° phase displacement. The harmonic spectrum in Fig. 3.3-4b indicates that i_a and $i_{\tilde{a}}$ contain the 5th and 7th harmonics, but these harmonics are canceled by the transformer and do not appear in i_A. It should be pointed out that the amplitudes of the fundamental component in i_a and i_A are not exactly the same. They would be identical if the voltage ratio of the transformer were equal to 2 instead of 2.05.

(3) THD and Power Factor. Figure 3.3-5 shows the calculated line current THD and input power factor versus the fundamental line current I_{A1}. The leakage inductance L_{lk} is typically 0.05 pu, while the line inductance L_s usually varies with the capacity and operating conditions of the power system. To investigate the effect of L_s, three typical values, $L_s = 0$, 0.05 pu and 0.1 pu, are selected. The THD of the line current i_A decreases with the increase of I_{A1} and L_s. Compared with the six-pulse rectifier, the 12-pulse rectifier can achieve a substantial reduction in the line

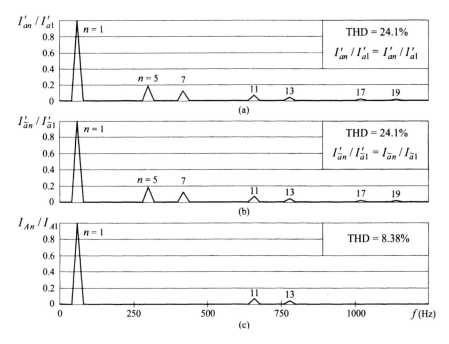

Figure 3.3-3 Harmonic spectrum of the current waveforms in Fig. 3.3-2.

current THD. Its input power factor PF is also improved thanks to the lower line current THD and higher displacement power factor.

Generally speaking, the THD of the line current in the 12-pulse rectifier does not meet the harmonic requirements set by IEEE Standard 519-1992. In practice, a line-side filter is normally required to reduce line current THD.

3.3.2 18-Pulse Series-Type Diode Rectifier

(1) Rectifier Configuration. The block diagram of an 18-pulse series-type diode rectifier is shown in Fig. 3.3-6. The rectifier has three units of identical six-pulse diode rectifiers fed by a phase shifting transformer. The sign "Z" enclosed by a circle represents a three-phase zigzag-connected winding, which provides a required phase displacement δ between the primary and secondary line-to-line voltages. The detailed analysis of the zigzag transformer is given in Chapter 5.

The 18-pulse rectifier is able to eliminate four dominant harmonics (the 5th, 7th, 11th, and 13th). This can be achieved by employing a phase-shifting transformer with a 20° phase displacement between any two adjacent secondary windings. The typical values of δ are 20°, 0°, and −20° for the top, middle, and bottom secondary windings, respectively. Other arrangements for δ are possible, such as $\delta = 0°$, 20°,

52 Chapter 3 Multipulse Diode Rectifiers

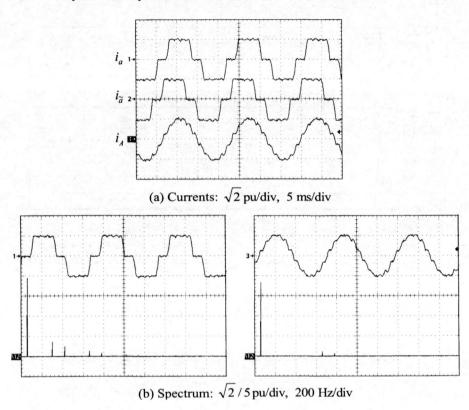

(a) Currents: $\sqrt{2}$ pu/div, 5 ms/div

(b) Spectrum: $\sqrt{2}/5$ pu/div, 200 Hz/div

Figure 3.3-4 Measured waveforms and harmonic spectrum of the 12-pulse series-type rectifier.

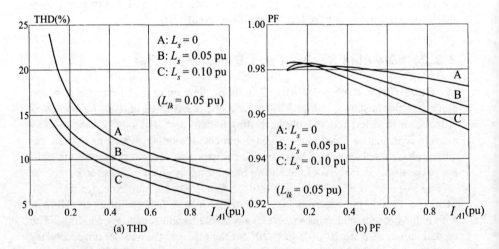

Figure 3.3-5 Line current THD and PF of the 12-pulse series-type diode rectifier.

3.3 Series-Type Multipulse Diode Rectifiers

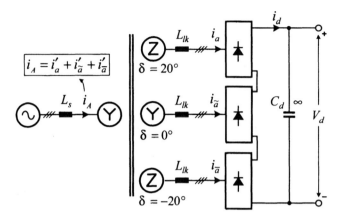

Figure 3.3-6 The 18-pulse series-type diode rectifier.

and 40°. The turns ratio of the transformer for the 18-pulse rectifier is usually selected such that the line-to-line voltage of each secondary winding is one third that of the primary winding.

(2) Waveforms. Assume that the 18-pulse diode rectifier in Fig. 3.3-6 operates under the rated conditions with $L_s = 0$ and $L_{lk} = 0.05$ pu. A set of simulated waveforms for the rectifier are shown in Fig. 3.3-7, where i'_a, $i'_{\tilde{a}}$, and $i'_{\bar{a}}$ are the primary current components referred from the secondary side of the transformer. The waveforms of these currents are all different due to the phase displacement of harmonic currents when they are transferred from the secondary to the primary winding. The waveforms of the secondary currents i_a, $i_{\tilde{a}}$, and $i_{\bar{a}}$ are not given in the figure, but they have the same shape as that of $i'_{\tilde{a}}$. The harmonic content of the primary and secondary line currents are given in Fig. 3.3-7c. The secondary line current has a THD of 23.7%, while the THD of the primary line current is only 3.06% due to the elimination of four dominant harmonics.

The waveforms measured from an 18-pulse diode rectifier under the rated operating conditions are shown in Fig. 3.3-8. The phase-shifting transformer has a leakage inductance of 0.05 pu and a voltage ratio of $V_{AB}/V_{ab} = V_{AB}/V_{\tilde{a}\tilde{b}} = V_{AB}/V_{\bar{a}\bar{b}} = 2.95$. The waveforms of the secondary line currents i_a, $i_{\tilde{a}}$, and $i_{\bar{a}}$ has a 20° phase displacement between each other. The harmonic spectrum illustrates that the primary line current i_A does not contain the 5th, 7th, 11th, or 13th harmonics and therefore is nearly sinusoidal.

(3) Line Current THD and Input Power Factor. Figure 3.3-9 shows the calculated primary line current THD and input power factor of the 18-pulse diode rectifier. Compared with the 12-pulse rectifier, the 18-pulse rectifier has lower line current THD and better power factor. For instance, when the 18-pulse oper-

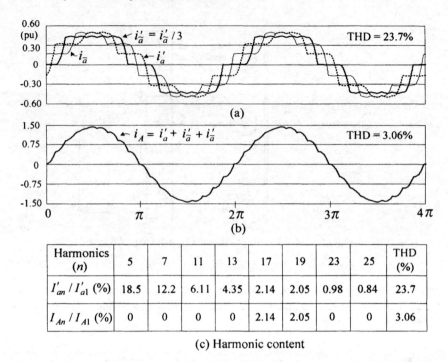

Figure 3.3-7 Current waveforms in the 18-pulse series-type rectifier ($L_s = 0$, $L_{lk} = 0.05$ pu, and $I_{A1} = 1.0$ pu).

ates under the rated load conditions ($I_{A1} = 1$ pu) with $L_s = 0.05$ pu, the THD of i_A is reduced from 6.4% of the 12-pulse rectifier to 2.3% and the power factor is slightly increased as well.

3.3.3 24-Pulse Series-Type Diode Rectifier

The configuration of a 24-pulse series-type diode rectifier is shown in Fig. 3.3-10, where a phase-shifting transformer is used to power four sets of six-pulse diode rectifiers. To eliminate six dominant current harmonics (the 5th, 7th, 11th, 13th, 17th, and 19th), the transformer should be arranged such that there is a 15° phase displacement between the voltages of any two adjacent secondary windings. The line-to-line voltage of each secondary winding is usually one fourth that of the primary winding.

Figure 3.3-11 shows the current waveforms in the 24-pulse diode rectifier operating under rated conditions, where i'_a, i''_a, i'''_a, and i''''_a are the primary currents referred from the secondary side of the transformer. Each of these currents has a THD of 24%. The primary line current i_A is virtually a sinusoid with only 1.49% total harmonic distortion.

3.3 Series-Type Multipulse Diode Rectifiers 55

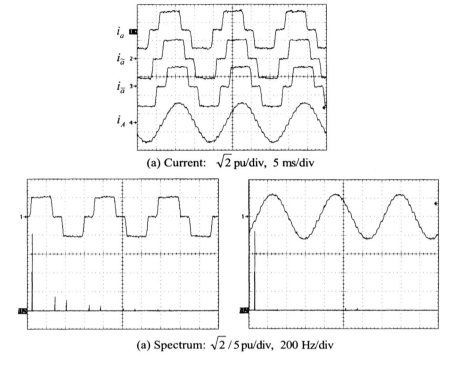

(a) Current: $\sqrt{2}$ pu/div, 5 ms/div

(a) Spectrum: $\sqrt{2}/5$ pu/div, 200 Hz/div

Figure 3.3-8 Waveforms and harmonic spectrum measured from an 18-pulse series-type rectifier.

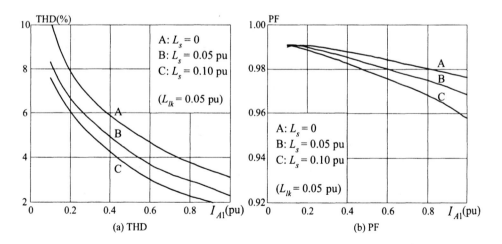

Figure 3.3-9 THD and PF of the 18-pulse series-type diode rectifier.

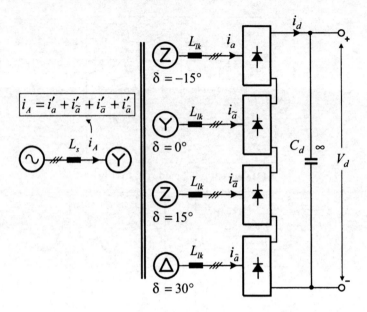

Figure 3.3-10 The 24-pulse series-type diode rectifier.

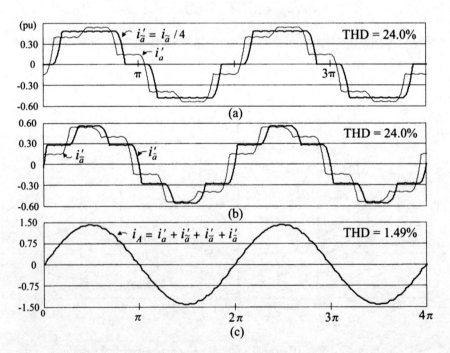

Figure 3.3-11 Current waveforms in the 24-pulse series-type rectifier ($L_s = 0$, $L_{lk} = 0.05$ pu and $I_{A1} = 1.0$ pu).

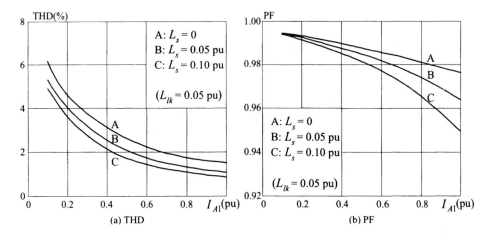

Figure 3.3-12 THD and PF of the 24-pulse series-type diode rectifier.

The calculated THD of the line current i_A and input power factor of the 24-pulse rectifier is shown in Fig. 3.3-12. It can be observed that the rectifier has an excellent THD profile, which meets the harmonic requirements specified by IEEE Standard 519-1992.

3.4 SEPARATE-TYPE MULTIPULSE DIODE RECTIFIERS

In the previous section, we have discussed series-type multipulse diode rectifiers, where the dc outputs of all the six-pulse diode rectifiers are connected in series. This section focuses on the separate-type multipulse rectifiers, where each of its six-pulse rectifiers feeds a separate dc load.

3.4.1 12-Pulse Separate-Type Diode Rectifier

The block diagram of a 12-pulse separate-type diode rectifier is shown in Fig. 3.4-1. The rectifier configuration is essentially the same as that of the 12-pulse series-type rectifier except that two separate dc loads are employed instead of a single dc load.

Figure 3.4-2 illustrates an application example of the 12-pulse separate-type rectifier as a front end for a cascaded H-bridge multilevel inverter-fed drive. The phase-shifting transformer has six secondary windings, of which three are wye-connected with $\delta = 0$ and the other three are delta-connected having a δ of 30°. Each of the secondary windings feeds a six-pulse diode rectifier. Since all wye-connected secondary windings are identical and so are the delta-connected windings, it is essentially a 12-pulse transformer. The six-pulse rectifiers provide isolated dc voltages to H-bridge inverters, whose outputs are connected in cascade, providing a three-phase ac voltage to the motor.

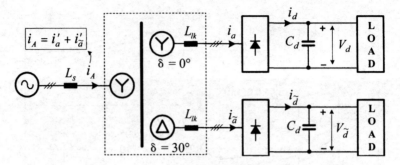

Figure 3.4-1 The 12-pulse separate-type diode rectifier.

Figure 3.4-3 shows the current waveforms of the 12-pulse separate-type rectifier operating with a rated line current. The waveform of the secondary line current i_a is similar to that of the "stand-alone" six-pulse rectifier. This is due to the fact that with the line inductance L_s assumed to be zero, the 12-pulse rectifier is essentially composed of two units of stand-alone six-pulse rectifiers. The dc currents in the two rectifiers, i_d and $i_{\tilde{d}}$, contain a higher ripple component compared with that in the 12-

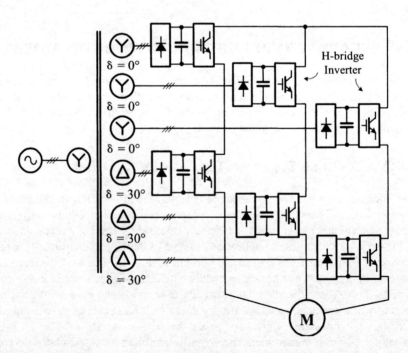

Figure 3.4-2 Application of the 12-pulse separate-type diode rectifier in a cascaded H-bridge multilevel inverter-fed drive.

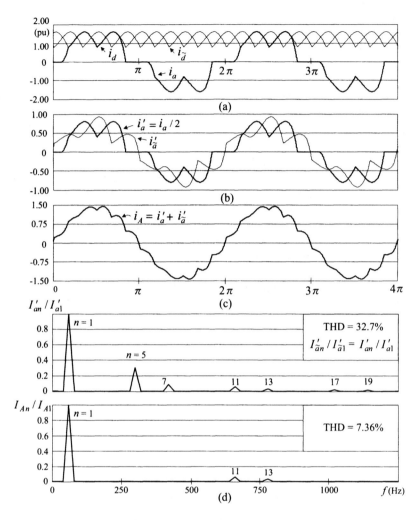

Figure 3.4-3 Current waveforms in the 12-pulse separate-type rectifier ($L_s = 0$, $L_{lk} = 0.05$ pu, and $I_{A1} = 1.0$ pu).

pulse series-type rectifier where the dc load sees the leakage inductances of the two secondary windings in series.

The currents i'_a and $i'_{\tilde{a}}$ in Fig. 3.4-3b are the secondary line currents i_a and $i_{\tilde{a}}$ referred to the primary side. For the reasons discussed earlier, the 5th and 7th harmonic currents in i'_a and $i'_{\tilde{a}}$ are out of phase and therefore are canceled in the primary winding of the transformer.

It is interesting to note that although the waveforms of i_a and $i_{\tilde{a}}$ differ significantly from those in the 12-pulse series-type rectifier, the primary line current i_A in both rectifiers has a similar waveform, and so does its THD. This is mainly due to the

60 Chapter 3 Multipulse Diode Rectifiers

12-pulse configuration, where the two most detrimental harmonics, the 5th and 7th, are eliminated. The remaining harmonics have less influence on the line current waveform and its THD.

Figure 3.4-4 shows the waveforms measured from a 12-pulse separate-type rectifier operating under the rated conditions. The phase shifting transformer has a leakage inductance of 0.045 pu and a voltage ratio of $V_{AB}/V_{ab} = V_{AB}/V_{\tilde{a}\tilde{b}} = 2.05$. The waveform of the secondary line currents, i_a and $i_{\tilde{a}}$, has two humps per half-cycle while the primary current i_A is close to sinusoidal due to the elimination of the 5th and 7th harmonics as shown in Fig. 3.4-4b.

The THD of the line current i_A in the 12-pulse separate-type rectifier shown in Fig. 3.4-5 is somewhat lower than that of the series type. This is mainly due to the differences in harmonic distribution. The secondary line currents in the separate-type rectifier contain higher 5th and 7th harmonics but lower 11th and 13th harmonics than those in the series type. When they are reflected to the primary side, the 5th and 7th harmonics are canceled, and thus the lower magnitude of the 11th and 13th harmonics makes a reduction in the line current THD.

The power factor profile is also different from that of the 12-pulse series-type rectifier. A notch occurs approximately at $I_{A1} = 0.22$ pu, which signifies a boundary be-

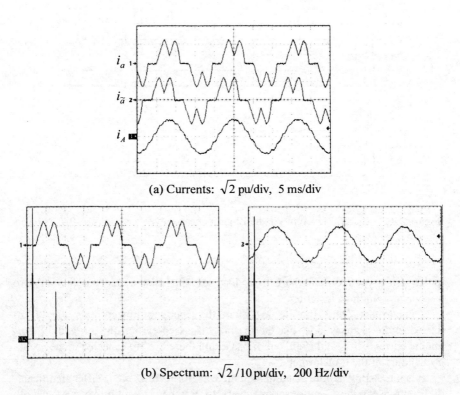

(a) Currents: $\sqrt{2}$ pu/div, 5 ms/div

(b) Spectrum: $\sqrt{2}/10$ pu/div, 200 Hz/div

Figure 3.4-4 Waveforms and harmonic spectrum measured from a 12-pulse separate-type rectifier.

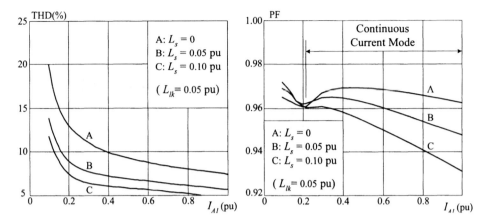

Figure 3.4-5 THD and PF of the 12-pulse separate-type diode rectifier.

tween continuous and discontinuous current operation of the rectifier. The discontinuous current operation normally does not occur in the series-type rectifiers since the dc load sees the leakage inductances of the secondary windings in series, making the dc current continuous over almost the full operation range. The power factor of the separate type is slightly lower than the series type. This is mainly due to the dc load connection, which affects the equivalent inductance seen by the utility supply.

3.4.2 18- and 24-Pulse Separate-Type Diode Rectifiers

The configuration of the 18-pulse separate-type diode rectifier is shown in Fig. 3.4-6. It is essentially the same as that of the 18-pulse series-type diode rectifier except for the dc side connection.

The waveforms of the 18-pulse separate-type rectifier are shown in Fig. 3.4-7. Due to the elimination of four dominant harmonics, the line current i_A is close to sinusoidal with THD of 3.05%.

Figure 3.4-8 shows the line current THD and input power factor of the 18- and 24-pulse separate-type rectifiers, respectively. In general, the THD profile of the separate-type rectifiers is somewhat better, and the power factor profile is slightly worse than their series-type counterparts.

3.5 SUMMARY

This chapter provides a comprehensive analysis on the multipulse diode rectifiers widely used in high-power medium voltage drives as front-end converters. The main issues discussed in the chapter are summarized below.

- **Systematic analysis on 12-, 18- and 24-pulse diode rectifiers.** The line current THD and input power factor of the multipulse rectifiers are analyzed.

62   Chapter 3   Multipulse Diode Rectifiers

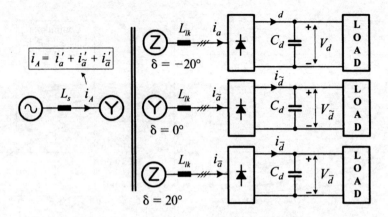

Figure 3.4-6   The 18-pulse separate-type diode rectifier.

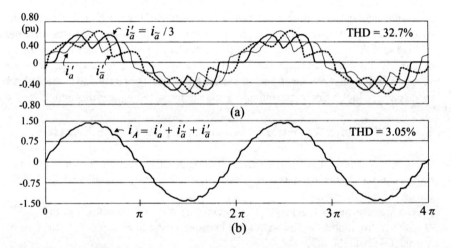

Figure 3.4-7   Current waveforms in the 18-pulse separate-type rectifier ($L_s = 0$, $L_{lk} = 0.05$ pu, and $I_{A1} = 1.0$ pu).

The line current THD of the 12-pulse diode rectifiers normally do not satisfy the harmonic requirements specified by IEEE Standard 519-1992. The 18-pulse rectifiers have a better harmonic profile, while the 24-pulse rectifiers provide excellent harmonic performance. The input power factor of the multipulse rectifiers is also analyzed. Rectifiers with more than 30 pulses are rarely used in practice mainly due to increased transformer costs and limited performance improvements.

- **Comparison between series- and separate-type rectifiers.** The multipulse rectifiers can be classified into series and separate types for use in various

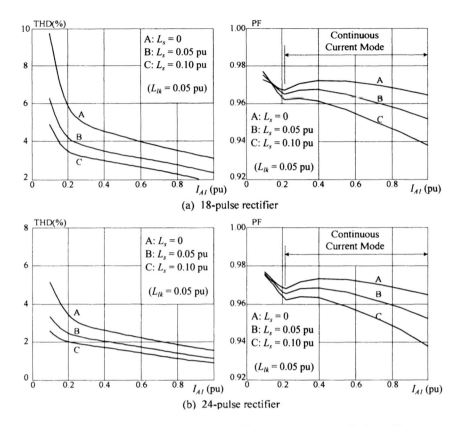

Figure 3.4-8 THD and PF of the 18- and 24-pulse separate-type diode rectifiers.

multilevel voltage source inverters. In general, the line current THD profile of the separate-type rectifiers is somewhat better, and the input power factor is slightly worse than their series-type counterparts.

REFERENCES

1. S. Malik and D. Kluge, ACS 1000—World's First Standard AC Drive for Medium Voltage Applications, *ABB Review*, No. 2, pp. 4–11, 1998.
2. E. A. Lewis, Power Converter Building Blocks for Multi-megawatt PWM VSI Drives, *IEE Seminar on PWM Medium Voltage Drives*, pp. 4/1–4/19, 2000.
3. Y. Shakweh, New Breed of Medium Voltage Converters, *IEE Power Engineering Journal*, pp. 12–20, 2000.
4. W. A. Hill and C. D. Harbourt, Performance of Medium Voltage Multilevel Inverters, *IEEE IAS Annual Meeting*, Vol. 2, pp. 1186–1192, 1999.
5. P. W. Hammond, A New Approach to Enhance Power Quality for Medium Voltage AC Drives, *IEEE Transactions on Industry Applications*, Vol. 33, No. 1, pp. 202–208, 1997.

6. B. Wu and F. DeWinter, Voltage Stress on Induction Motor in Medium Voltage (2300 V to 6900 V) PWM GTO CSI Drives, *IEEE Transactions on Power Electronics,* Vol. 12, No. 2, pp. 213–220, 1997.
7. J. Das, and R. H. Osman, Grounding of AC and DC low-voltage and Medium Voltage Drive Systems, *IEEE Transactions on Industry Applications,* Vol. 34, No. 1 pp. 205–216, 1998.

Chapter 4

Multipulse SCR Rectifiers

4.1 INTRODUCTION

The multipulse diode rectifiers presented in the previous chapter are normally used in voltage source inverter (VSI)-fed drives, while the multipulse SCR rectifiers to be discussed in this chapter are mainly for current source inverter (CSI)-based drives. The SCR rectifier provides an adjustable dc current for the CSI which converts the dc current to a three-phase PWM ac current with variable frequencies.

This chapter starts with an overview of six-pulse SCR rectifier, which is the building block for the multipulse SCR rectifiers, followed by an analysis on 12-, 18-, and 24-pulse rectifiers. The line current THD and input power factor of these rectifiers are investigated, and the results are summarized in a graphic format.

4.2 SIX-PULSE SCR RECTIFIER

Figure 4.2-1 shows a simplified circuit diagram for the six-pulse SCR rectifier, where RC snubber circuits for the SCR devices are omitted. The line inductance L_s represents the total inductance between the utility supply and the rectifier, including the equivalent inductance of the supply, the total leakage inductance of isolation transformer if any, and the inductance of a three-phase line reactor that is often added to the system for the reduction of line current THD. On the dc side of the rectifier, a dc choke L_d is used to smooth the dc current. The choke is normally constructed with a single magnetic core and two coils, one coil in the positive dc bus and the other in the negative bus. Such an arrangement is preferable in medium-voltage drives since it helps to reduce the common-mode voltage imposed on the motor without increasing the manufacturing cost of the choke [1].

To simplify the analysis, it is assumed that the inductance of the dc choke L_d is sufficiently high such that the dc current I_d is ripple-free. The dc choke and the load can then be replaced by an adjustable dc current source as shown in Fig. 4.2-1b.

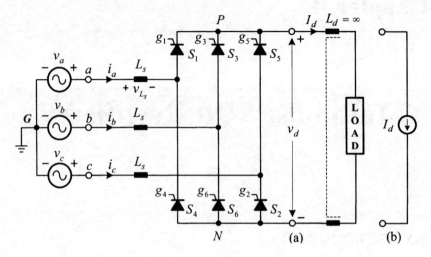

Figure 4.2-1 Simplified circuit diagram of a six-pulse SCR rectifier.

4.2.1 Idealized Six-Pulse Rectifier

Let's consider an idealized six-pulse SCR rectifier, where the line inductance L_s in Fig. 4.2-1 is assumed to be zero. Figure 4.2-2 shows typical waveforms of the rectifier, where v_a, v_b, and v_c are the phase voltages of the utility supply, i_{g1} to i_{g6} are the gate signals for SCR switches S_1 to S_6, and α is the **firing angle** of the SCRs, respectively.

During interval I ($\pi/6 + \alpha \leq \omega t < \pi/2 + \alpha$), v_a is higher than the other two phase voltages (v_b and v_c), making S_1 forward-biased. When S_1 is fired at $\omega t = \pi/6 + \alpha$ by its gate signal i_{g1}, it is turned on. The positive dc bus voltage v_P with respect to ground G is equal to v_a. Assuming that S_6 was conducting prior to the turn-on of S_1, it continues to conduct until the end of interval I, during which the negative bus voltage v_N is equal to v_b. The dc output voltage can be found from $v_d = v_P - v_N = v_{ab}$. The dc current I_d flows from v_a to v_b through S_1, the load, and S_6. The three-phase line currents are $i_a = I_d$, $i_b = -I_d$, and $i_c = 0$ as shown in Fig. 4.2-2.

During interval II ($\pi/2 + \alpha \leq \omega t < 5\pi/6 + \alpha$), v_c is lower than the other two phase voltages (v_a and v_b), making S_2 forward-biased. At the moment the gate signal i_{g2} arrives, S_2 is switched on. The conduction of S_2 makes S_6 reverse-biased, forcing it to turn off. The dc current I_d is then commutated from S_6 to S_2, which leads to $i_b = 0$ and $i_c = -I_d$. The commutation process in this case completes instantly due to the absence of the line inductance. The dc output voltage is given by $v_d = v_P - v_N = v_{ac}$. Following the same procedure, all the current and voltage waveforms in other intervals can be obtained.

The average dc output voltage can be determined by

$$V_d = \frac{\text{area } A_1}{\pi/3} = \frac{1}{\pi/3} \int_{\pi/6+\alpha}^{\pi/2+\alpha} v_{ab} d(\omega t) = \frac{3\sqrt{2}}{\pi} V_{LL} \cos \alpha = 1.35 V_{LL} \cos \alpha \quad (4.2\text{-}1)$$

4.2 Six-Pulse SCR Rectifier

where $v_{ab} = \sqrt{2}V_{LL}\sin(\omega t + \pi/6)$. The equation illustrates that the rectifier dc output voltage V_d is positive when the firing angle α is less than $\pi/2$ and becomes negative for an α greater than $\pi/2$. However, the dc current I_d is always positive, irrelevant to the polarity of the dc output voltage.

When the rectifier produces a positive dc voltage, the power is delivered from the supply to the load. With a negative dc voltage, the rectifier operates in an **inverting mode**, and the power is fed from the load back to the supply. This often takes place in a CSI drive during rapid speed deceleration where the kinetic energy of the rotor and its mechanical load is converted to the electrical energy by the inverter and then sent back to the power supply by the SCR rectifier for fast **dynamic braking**. The power flow in the SCR rectifier is, therefore, **bidirectional**, which also enables the CSI drive to operate in four quadrants, an important feature provided by the SCR rectifier.

The line current i_a in Fig. 4.2-2 can be expressed in a Fourier series as

$$i_a = \frac{2\sqrt{3}}{\pi} I_d \bigg(\sin(\omega t - \phi_1) - \frac{1}{5}\sin 5(\omega t - \phi_1) - \frac{1}{7}\sin 7(\omega t - \phi_1) + \frac{1}{11}\sin 11(\omega t - \phi_1)$$

$$+ \frac{1}{13}\sin 13(\omega t - \phi_1) - \frac{1}{17}\sin 17(\omega t - \phi_1) - \frac{1}{19}\sin 19(\omega t - \phi_1) + \cdots \bigg) \quad (4.2\text{-}2)$$

where ϕ_1 is the phase angle between the supply voltage v_a and the fundamental-frequency line current i_{a1}.

The rms value of i_a can be calculated by

$$I_a = \left(\frac{1}{2\pi} \int_0^{2\pi} (i_a)^2 d(\omega t) \right)^{1/2} = \left(\frac{1}{2\pi} \left(\int_{\frac{\pi}{6}+\alpha}^{\frac{5\pi}{6}+\alpha} (I_d)^2 d(\omega t) + \int_{\frac{7\pi}{6}+\alpha}^{\frac{11\pi}{6}+\alpha} (-I_d)^2 d(\omega t) \right) \right)^{1/2}$$

$$= \sqrt{\frac{2}{3}} I_d = 0.816 I_d \quad (4.2\text{-}3)$$

from which the total harmonic distortion for the line current i_a is

$$\text{THD} = \frac{\sqrt{I_a^2 - I_{a1}^2}}{I_{a1}} = \frac{\sqrt{(0.816 I_d)^2 - (0.78 I_d)^2}}{0.78 I_d} = 0.311 \quad (4.2\text{-}4)$$

where I_{a1} is the rms value of i_{a1}.

To find the displacement power factor (DPF), we can refer to β_1 and β_2 in Fig. 4.2-2. Since β_1 is fixed to $\pi/6$ and β_2 is equal to $\pi/6 + \alpha$, the displacement power factor angle is

$$\phi_1 = \beta_2 - \beta_1 = \alpha$$

from which

$$\text{DPF} = \cos \phi_1 = \cos \alpha \quad (4.2\text{-}5)$$

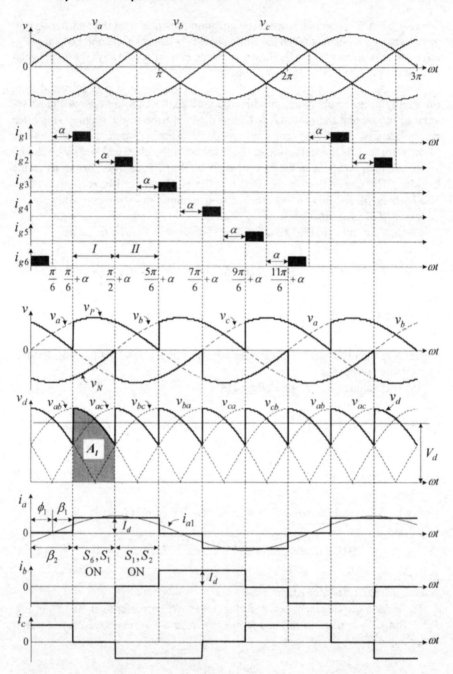

Figure 4.2-2 Waveforms of the idealized six-pulse SCR rectifier operating at $\alpha = 30°$.

The overall power factor for the six-pulse SCR rectifier can be obtained from

$$PF = DPF \times DF = \frac{\cos \phi_1}{\sqrt{1 + THD^2}} = 0.955 \cos \alpha \qquad (4.2\text{-}6)$$

where DF is the distortion factor defined in Chapter 3.

Figure 4.2-3 shows the voltage waveforms of the rectifier with various firing angles. The average dc output voltage V_d is positive at $\alpha = 45°$, falls to zero at α

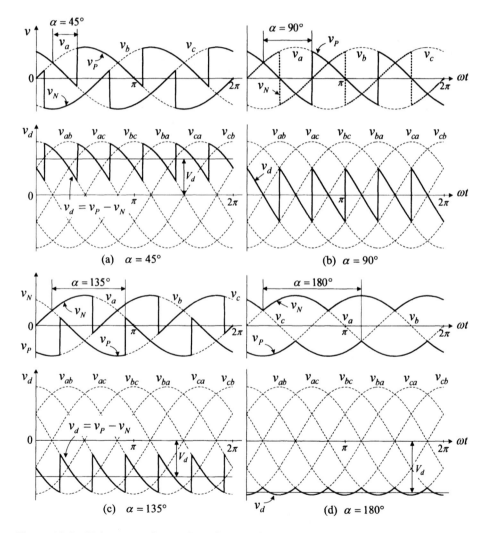

Figure 4.2-3 Voltage waveforms of the idealized six-pulse SCR rectifier operating at various firing angles.

= 90°, and becomes negative when $\alpha = 135°$. It reaches its maximum negative value at $\alpha = 180°$. In a practical rectifier where the line inductance L_s is present, the firing angle α should be less than 180° to prevent SCR commutation failure [2].

4.2.2 Effect of Line Inductance

With the presence of the line inductance L_s, the commutation of the SCR devices will not complete instantly. Consider a case where the dc output current I_d is commutated from S_5 to S_1 as shown in Fig. 4.2-4. Assuming that S_5 and S_6 are conducting prior to the turn-on of S_1, the dc current flows through both devices. The commutation process is initiated by turning S_1 on at α. At the moment the incoming device S_1 is gated on, its current i_a starts to rise from zero, but cannot jump to I_d instantly due to the line inductance L_s. In the meantime, the current i_c in the outgoing device S_5 starts to decrease since $i_c = I_d - i_a$. As a result, three SCR devices, S_1, S_5 and S_6, conduct simultaneously. The commutation completes at the end of the commutation interval γ, at which the current i_a in S_1 reaches I_d whereas the current i_c in S_5 falls to zero.

The commutation causes a reduction in the average dc voltage V_d. Since both S_1 and S_5 conduct simultaneously during the γ interval, the positive bus voltage v_P with respect to ground G can be expressed as

$$v_P = -L_s \frac{di_a}{dt} + v_a = -L_s \frac{di_c}{dt} + v_c \qquad (4.2\text{-}7)$$

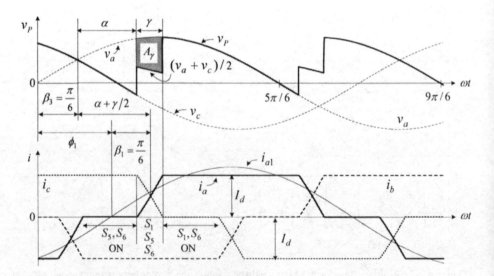

Figure 4.2-4 Voltage and current waveforms during commutation ($\alpha = 45°$).

4.2 Six-Pulse SCR Rectifier

from which

$$v_P = \frac{v_a + v_c}{2} - \frac{L_s}{2}\left(\frac{di_a}{dt} + \frac{di_c}{dt}\right) \qquad (4.2\text{-}8)$$

Since $i_a + i_c = I_d = $ const, we have

$$\frac{di_a}{dt} + \frac{di_c}{dt} = 0 \qquad (4.2\text{-}9)$$

Substituting (4.2-9) into (4.2-8) leads to

$$v_P = \frac{v_a + v_c}{2} \qquad (4.2\text{-}10)$$

The waveform of v_P during the γ interval is also shown in Fig. 4.2-4. The shaded area A_γ, representing the amount of voltage reduction caused by the commutation, can be found from

$$A_\gamma = \int_{\frac{\pi}{6}+\alpha}^{\frac{\pi}{6}+\alpha+\gamma} (v_a - v_P) d(\omega t) \qquad (4.2\text{-}11)$$

where

$$v_a - v_P = L_s(di_a/dt) \qquad (4.2\text{-}12)$$

Substituting (4.2-12) into (4.2-11) yields

$$A_\gamma = \int_0^{I_d} \omega L_s di_a = \omega L_s I_d \qquad (4.2\text{-}13)$$

The average dc voltage loss ΔV can then be calculated by

$$\Delta V = \frac{A_\gamma}{\pi/3} = \frac{3\omega L_s}{\pi} I_d \qquad (4.2\text{-}14)$$

Taking the effect of the line inductance L_s into account, the average dc output voltage of the six-pulse SCR rectifier is

$$V_d = 1.35 V_{LL} \cos\alpha - \frac{3\omega L_s}{\pi} I_d \qquad (4.2\text{-}15)$$

The commutation angle γ can be derived from (4.2-11):

$$\gamma = \cos^{-1}\left(\cos\alpha - \frac{\sqrt{2}\omega L_s}{V_{LL}} I_d\right) - \alpha \qquad (4.2\text{-}16)$$

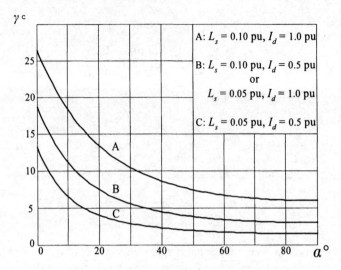

Figure 4.2-5 Commutation angle γ versus firing angle α.

Figure 4.2-5 shows the relationship between the commutation angle γ and the firing angle α. For a given α, the lower the value for L_s and I_d, the smaller the commutation angle γ is. The input power factor is affected by the line inductance L_s as well. Assuming that i_a and i_c shown in Fig. 4.2-4 varies linearly over time during the commutation interval, β_1 is equal to $\pi/6$. The displacement power factor angle ϕ_1 can be calculated by

$$\phi_1 = \beta_3 + (\alpha + \gamma/2) - \beta_1 = \alpha + \gamma/2 \qquad (4.2\text{-}17)$$

from which

$$\text{DPF} = \cos\phi_1 = \cos(\alpha + \gamma/2) \qquad (4.2\text{-}18)$$

The overall power factor of the rectifier can be determined by

$$\text{PF} = \text{DPF} \times \text{DF} = \frac{\cos(\alpha + \gamma/2)}{\sqrt{1 + \text{THD}^2}} \qquad (4.2\text{-}19)$$

4.2.3 Power Factor and THD

Figure 4.2-6 shows the simulated waveforms for the line current i_a when the rectifier operates with the rated line current ($I_{a1} = 1$ pu). The line inductance L_s is assumed to be 0.05 pu, and the firing angle α is 0° in Fig. 4.2-6a and 30° in Fig. 4.2-6b, respectively. It is interesting to note that waveform of i_a during the γ interval varies with α. It rises nonlinearly when $\alpha = 0°$ and looks somewhat like lin-

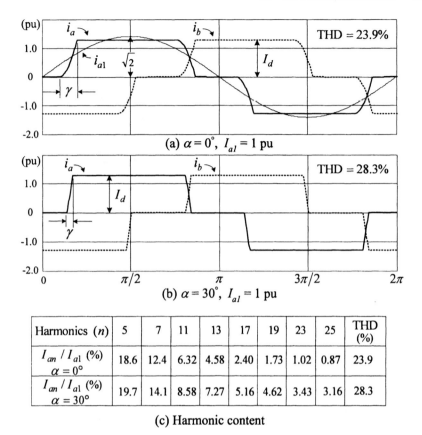

Figure 4.2-6 Line current waveforms in the six-pulse SCR rectifier with $L_s = 0.05$ pu.

ear for $\alpha = 30°$. This is because the line current i_a is a function of α during commutation, given by

$$i_a = \frac{V_{LL}}{\sqrt{2}\omega L_s}(\cos\alpha - \cos(\omega t + \alpha)), \qquad 0 \leq \omega t \leq \gamma \qquad (4.2\text{-}20)$$

Figure 4.2-6c shows the line current harmonic content for the six-pulse SCR rectifier. Its THD is more than 20%, which is not acceptable in practice, especially when the rectifier is for high-power applications.

Figure 4.2-7 shows the line current THD versus I_{a1} with L_s and α as parameters. The THD reduces with the increase of I_{a1} and L_s as shown in Fig. 4.2-7a. It also decreases with the firing angle α as illustrated in Fig. 4.2-7b.

Figure 4.2-8 shows the input power factor profile of the six-pulse SCR rectifier as a function of I_{a1} and α. The power factor varies slightly with the line current I_{a1}.

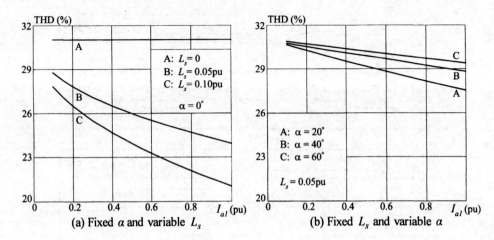

Figure 4.2-7 Line current THD of the six-pulse SCR rectifier.

However, it reduces substantially with large values of α. This is, in fact, the main drawback of the SCR rectifier.

4.3 12-PULSE SCR RECTIFIER

The block diagram of a 12-pulse SCR rectifier is shown in Fig. 4.3-1. It is composed of a phase-shifting transformer and two identical six-pulse SCR rectifiers. The transformer has two secondary windings, one connected in wye and the other in delta. The line-to-line voltage of the secondary windings is normally half of its pri-

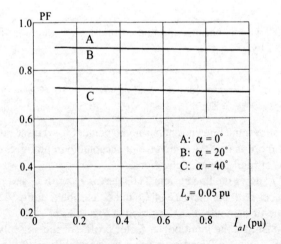

Figure 4.2-8 Power factor of the six-pulse SCR rectifier.

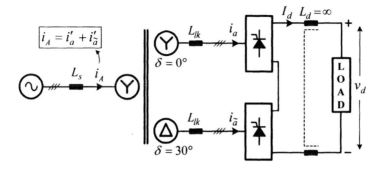

Figure 4.3-1 Block diagram of a 12-pulse SCR rectifier.

mary line-to-line voltage. The dc outputs of the two SCR rectifiers are connected in series for a single dc load. The dc choke L_d is assumed to be sufficiently large and the resultant dc current I_d is ripple-free.

As shown in Fig. 4.3-2, the 12-pulse SCR rectifier can be used as a front end for a CSI-fed drive. The inverter converts the dc current I_d to a three-phase PWM current i_w. The magnitude of i_w is proportional to I_d, and thus it can be adjusted by the rectifier through firing angle control. The details of the CSI drive will be discussed in the later chapters.

4.3.1 Idealized 12-Pulse Rectifier

Consider an idealized 12-pulse rectifier where the line inductance L_s and the total leakage inductance L_{lk} of the transformer are assumed to be zero. The current waveforms in the rectifier are shown in Fig. 4.3-3, where i_a and $i_{\tilde{a}}$ are the secondary line

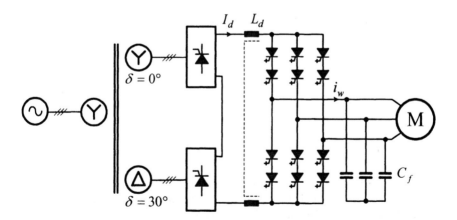

Figure 4.3-2 A CSI-fed drive using the 12-pulse SCR rectifier as a front end.

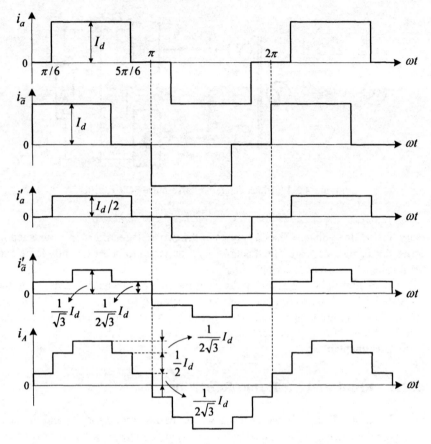

Figure 4.3-3 Current waveforms of the 12-pulse SCR rectifier ($L_s = L_{lk} = 0$).

currents, i'_a and $i'_{\tilde{a}}$ are the primary currents referred from the secondary side, and i_A is the primary line current given by $i_A = i'_a + i'_{\tilde{a}}$, respectively.

The secondary line current i_a can be expressed as

$$i_a = \frac{2\sqrt{3}}{\pi} I_d \left(\sin \omega t - \frac{1}{5} \sin 5\omega t - \frac{1}{7} \sin 7\omega t + \frac{1}{11} \sin 11\omega t + \frac{1}{13} \sin 13\omega t \right.$$

$$\left. - \frac{1}{17} \sin 17\omega t - \frac{1}{19} \sin 19\omega t + \cdots \right) \quad (4.3\text{-}1)$$

where $\omega = 2\pi f_1$ is the angular frequency of the supply voltage. Since the waveform of i_a is of half-wave symmetry, it does not contain any even-order harmonics. In addition, i_a does not contain any triplen harmonics either due to the balanced three-phase system.

4.3 12-Pulse SCR Rectifier

The other secondary current $i_{\tilde{a}}$ leads i_a by 30°, and its Fourier expression is

$$i_{\tilde{a}} = \frac{2\sqrt{3}}{\pi} I_d \bigg(\sin(\omega t + 30°) - \frac{1}{5}\sin 5(\omega t + 30°) - \frac{1}{7}\sin 7(\omega t + 30°)$$

$$+ \frac{1}{11}\sin 11(\omega t + 30°) + \frac{1}{13}\sin 13(\omega t + 30°) - \frac{1}{17}\sin 17(\omega t + 30°)$$

$$- \frac{1}{19}\sin 19(\omega t + 30°) + \cdots \bigg) \tag{4.3-2}$$

The waveform for the referred current i'_a in Fig. 4.3-3 is identical to i_a except that its magnitude is halved due to the turns ratio of the Y/Y-connected windings. The current i'_a can be expressed in Fourier series as

$$i'_a = \frac{\sqrt{3}}{\pi} I_d \bigg\{ \sin \omega t - \frac{1}{5}\sin 5\omega t - \frac{1}{7}\sin 7\omega t + \frac{1}{11}\sin 11\omega t + \frac{1}{13}\sin 13\omega t$$

$$- \frac{1}{17}\sin 17\omega t - \frac{1}{19}\sin 19\omega t + \cdots \bigg\} \tag{4.3-3}$$

When the current $i_{\tilde{a}}$ is referred to the primary side, the phase angles of some harmonic currents are altered due to the Y/Δ-connected windings. As a result, the referred current $i'_{\tilde{a}}$ does not keep the same wave shape as $i_{\tilde{a}}$. The Fourier expression for $i'_{\tilde{a}}$ is

$$i'_{\tilde{a}} = \frac{\sqrt{3}}{\pi} I_d \bigg\{ \sin \omega t + \frac{1}{5}\sin 5\omega t + \frac{1}{7}\sin 7\omega t + \frac{1}{11}\sin 11\omega t + \frac{1}{13}\sin 13\omega t$$

$$+ \frac{1}{17}\sin 17\omega t + \frac{1}{19}\sin 19\omega t + \cdots \bigg\} \tag{4.3-4}$$

The line current i_A can be found from

$$i_A = i'_a + i'_{\tilde{a}} = \frac{2\sqrt{3}}{\pi} I_d \bigg\{ \sin \omega t + \frac{1}{11}\sin 11\omega t + \frac{1}{13}\sin 13\omega t + \frac{1}{23}\sin 23\omega t$$

$$+ \frac{1}{25}\sin 25\omega t + \cdots \bigg\} \tag{4.3-5}$$

where the two dominant current harmonics, the 5th and 7th, are canceled in addition to the 17th and 19th.

The THD of the secondary and primary line currents i_a and i_A can be determined by

$$\text{THD}(i_a) = \frac{\sqrt{I_a^2 - I_{a1}^2}}{I_{a1}} = \frac{(I_{a5}^2 + I_{a7}^2 + I_{a11}^2 + I_{a13}^2 + \cdots)^{1/2}}{I_{a1}} = 31.1\% \tag{4.3-6}$$

and

$$\text{THD}(i_A) = \frac{\sqrt{I_A^2 - I_{A1}^2}}{I_{A1}} = \frac{(I_{A11}^2 + I_{A13}^2 + I_{A23}^2 + I_{A25}^2 + \cdots)^{1/2}}{I_{A1}} = 15.3\% \quad (4.3\text{-}7)$$

The THD of the primary line current i_A in the idealized 12-pulse rectifier is reduced approximately by 50% compared with that of the secondary line current i_a.

4.3.2 Effect of Line and Leakage Inductances

Figure 4.3-4 shows typical current waveforms for the 12-pulse rectifier taking into account the transformer leakage inductance L_{lk}. The rectifier operates under the condition of $\alpha = 0°$, $I_{A1} = 1$ pu, $L_s = 0$ and $L_{lk} = 0.05$ pu. The waveform for the secondary line current i_a is close to a trapezoid and contains the 5th and 7th harmonics with a magnitude of 18.8% and 12.7%, respectively. However, these two harmonics are canceled by the phase-shifting transformer, and thus they do not appear in the primary line current i_A. Due to the effect of the leakage inductance, the THD of i_A is reduced from 15.3% in the idealized rectifier to 8.61%.

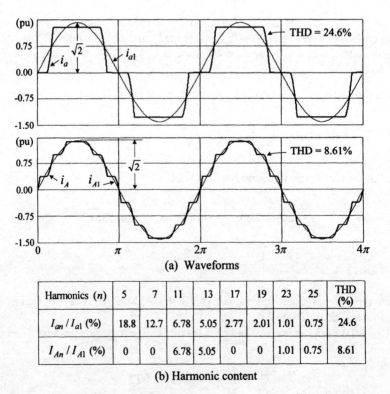

(a) Waveforms

Harmonics (n)	5	7	11	13	17	19	23	25	THD (%)
I_{an}/I_{a1} (%)	18.8	12.7	6.78	5.05	2.77	2.01	1.01	0.75	24.6
I_{An}/I_{A1} (%)	0	0	6.78	5.05	0	0	1.01	0.75	8.61

(b) Harmonic content

Figure 4.3-4 Typical current waveforms and harmonic contents of the 12-pulse SCR rectifier with $L_s = 0$ and $L_{lk} = 0.05$ pu.

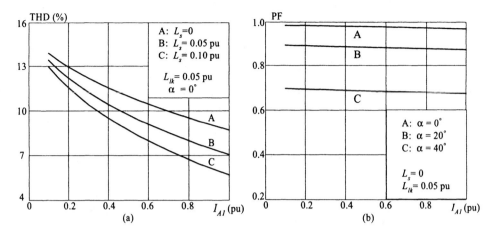

Figure 4.3-5 Primary line current THD and input PF of the 12-pulse SCR rectifier.

4.3.3 THD and PF

The THD of the primary line current i_A as a function of I_{A1} and L_s is illustrated in Fig. 4.3-5a. Compared with the six-pulse SCR rectifier, the 12-pulse rectifier has a much better THD profile. However, it generally does not meet the harmonic guidelines set by IEEE Standard 519-1992. The input power factor of the rectifier varies greatly with the firing angle as shown in Fig. 4.3-5b.

4.4 18- AND 24-PULSE SCR RECTIFIERS

The block diagram of an 18-pulse SCR rectifier is depicted in Fig. 4.4-1. Similar to the 18-pulse diode rectifiers, the rectifier employs a phase-shifting transformer with three secondary windings feeding three identical six-pulse SCR rectifiers. The configuration of the 24-pulse SCR rectifier can be easily derived and thus is not shown.

Figure 4.4-2 shows the typical current waveforms of the 18-pulse SCR rectifier operating under the condition of $\alpha = 0°$, $I_{A1} = 1$ pu, $L_s = 0$, and $L_{lk} = 0.05$ pu, where i'_a, $i'_{\tilde{a}}$ and $i'_{\bar{a}}$ are the primary currents referred from the transformer secondary side. All these currents have the same THD of 24.6%, although their waveforms are all different. The primary line current i_A does not contain the 5th, 7th, 11th, or 13th harmonics, resulting in a nearly sinusoidal waveform with a THD of only 3.54%.

Figure 4.4-3 shows the primary line current THD for the 18- and 24-pulse SCR rectifiers versus I_{A1} with the line inductance L_s as a parameter. As expected, the 18-pulse rectifier has better line current THD profile than the 12-pulse SCR rectifier while the 24-pulse rectifier is superior to the 18-pulse rectifier. The input power factor of the 18- and 24-pulse rectifiers is similar to that of the 12-pulse and therefore is not presented.

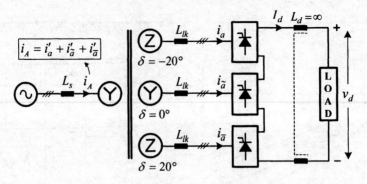

Figure 4.4-1 Block diagram of an 18-pulse SCR rectifier.

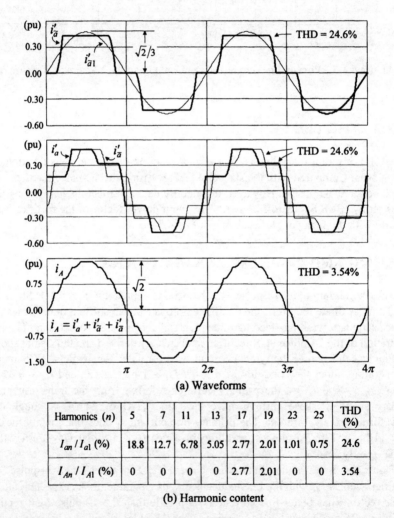

(a) Waveforms

Harmonics (n)	5	7	11	13	17	19	23	25	THD (%)
I_{an}/I_{a1} (%)	18.8	12.7	6.78	5.05	2.77	2.01	1.01	0.75	24.6
I_{An}/I_{A1} (%)	0	0	0	0	2.77	2.01	0	0	3.54

(b) Harmonic content

Figure 4.4-2 Current waveforms and harmonic contents of the 18-pulse SCR rectifier with $L_s = 0$ and $L_{lk} = 0.05$ pu.

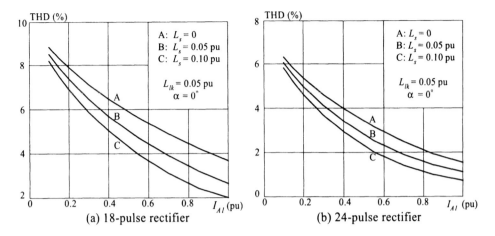

Figure 4.4-3 THD of the primary line current i_A in the 18- and 24-pulse SCR rectifiers.

4.5 SUMMARY

In this chapter, the operation of the six-pulse SCR rectifier is introduced and its performance is analyzed. The six-pulse rectifier is the building block for the multipulse SCR rectifiers, and therefore it is discussed in detail. The line current THD of the 12-pulse SCR rectifier normally does not satisfy the harmonic guidelines set by IEEE Standard 519-1992. The 18-pulse SCR rectifier has a better line current harmonic profile, while the 24-pulse rectifier provides a superior harmonic performance. The input power factor of the SCR rectifiers varies with the firing angle, which is the major disadvantage of the rectifiers.

The multipulse SCR rectifiers are naturally suited for use in medium-voltage CSI-fed drives. Over the last decade, the 18-pulse SCR rectifier has been a preferred choice for the CSI drive as a front end due to its good performance to price ratio. However, the SCR rectifier starts to be replaced by PWM GCT current source rectifiers for higher input power factor and better dynamic performance.

REFERENCES

1. B. Wu and F. DeWinter, Voltage Stress on Induction Motor in Medium Voltage (2300 V to 6900 V) PWM GTO CSI Drives, *IEEE Transactions on Power Electronics,* Vol. 12, No. 2, pp. 213–220, 1997.
2. N. Mohan, T. Undeland and W. P. Bobbins, *Power Electronic—Converters, Applications and Design,* 3rd edition, John Wiley & Sons, New York, 2003.

Chapter 5

Phase-Shifting Transformers

5.1 INTRODUCTION

The phase-shifting transformer is an indispensable device in multipulse diode/SCR rectifiers. It provides three main functions: (a) a required phase displacement between the primary and secondary line-to-line voltages for harmonic cancellation, (b) a proper secondary voltage, and (c) an electric isolation between the rectifier and the utility supply. According to the winding arrangements, the transformers can be classified into Y/Z and Δ/Z configurations, where the primary winding can be connected in wye (Y) or delta (Δ) while the secondary windings are normally in zigzag (Z) connection. Both configurations can be equally used in the multipulse rectifiers.

In this chapter, a number of issues concerning the phase-shifting transformer are addressed, including the configuration of the transformer, the design of turns ratios, and the principle of harmonic current cancellation.

5.2 Y/Z PHASE-SHIFTING TRANSFORMERS

Depending on winding connections, the line-to-line voltage of the transformer secondary winding may lead or lag its primary voltage by a phase angle δ. The Y/Z-1 transformers to be presented below provide a leading phase angle, while the Y/Z-2 transformers generate a lagging angle.

5.2.1 Y/Z-1 Transformers

Figure 5.2-1 shows a Y/Z-1 phase-shifting transformer and its phasor diagram. The primary winding is connected in wye with N_1 turns per phase. The secondary winding is composed of two sets of coils having N_2 and N_3 turns per phase. The N_2 coils are connected in delta and then in series with the N_3 coils. Such an arrangement is known as **zigzag** or **extended-delta** connection. As shown in its phasor diagram, the transformer can produce a phase shifting angle δ, defined by

$$\delta = \angle \overline{V}_{ab} - \angle \overline{V}_{AB} \qquad (5.2\text{-}1)$$

High-Power Converters and ac Drives. By Bin Wu
© 2006 The Institute of Electrical and Electronics Engineers, Inc.

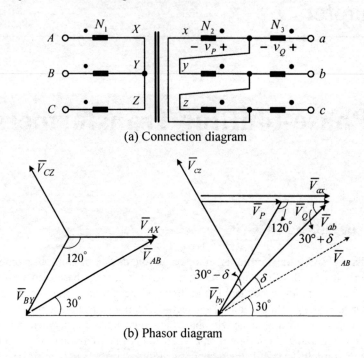

Figure 5.2-1 Y/Z-1 phase-shifting transformer.

where $\overline{V}_{AB}$ and $\overline{V}_{ab}$ are the phasors for the primary and secondary line-to-line voltages v_{AB} and v_{ab}, respectively. To determine the turns ratio for the transformer, let's consider a triangle composed of $\overline{V}_Q$, $\overline{V}_{by}$, and $\overline{V}_{ab}$ in the phasor diagram, from which

$$\frac{V_Q}{\sin(30° - \delta)} = \frac{V_{by}}{\sin(30° + \delta)}, \qquad 0° \leq \delta \leq 30° \qquad (5.2\text{-}2)$$

where V_Q is the rms voltage across the N_3 coil and V_{by} is the rms phase voltage between notes b and y. Since V_{by} is equal to V_{ax} in a balanced three-phase system, (5.2-2) can be rewritten as

$$\frac{V_Q}{V_{ax}} = \frac{\sin(30° - \delta)}{\sin(30° + \delta)} \qquad (5.2\text{-}3)$$

from which the turns ratio of the secondary coils is

$$\frac{N_3}{N_2 + N_3} = \frac{V_Q}{V_{ax}} = \frac{\sin(30° - \delta)}{\sin(30° + \delta)} \qquad (5.2\text{-}4)$$

For a given value of δ, the ratio of N_3 to $(N_2 + N_3)$ can be determined.

Similarly, the following relationship can be derived:

$$\frac{V_{ab}}{\sin 120°} = \frac{V_{by}}{\sin(30° + \delta)} \tag{5.2-5}$$

from which

$$V_{ax} = V_{by} = \frac{2}{\sqrt{3}} \sin(30° + \delta) V_{ab} \tag{5.2-6}$$

The turns ratio of the transformer is defined by

$$\frac{N_1}{N_2 + N_3} = \frac{V_{AX}}{V_{ax}} \tag{5.2-7}$$

Substituting (5.2-6) into (5.2-7) yields

$$\frac{N_1}{N_2 + N_3} = \frac{1}{2\sin(30° + \delta)} \frac{V_{AB}}{V_{ab}} \tag{5.2-8}$$

where $V_{AB} = \sqrt{3} V_{AX}$.

Let's now examine two extreme cases. Assuming that N_2 is reduced to zero, the secondary winding in Fig. 5.2-1 becomes wye-connected, and thus V_{ab} is in phase with V_{AB}, leading to $\delta = 0°$. Alternatively, if $N_3 = 0$, the secondary winding becomes delta-connected, resulting in $\delta = 30°$. Therefore, the phase-shifting angle δ for the Y/Z-1 transformer is in the range of $0°$ to $30°$.

5.2.2 Y/Z-2 Transformers

The configuration of a Y/Z-2 phase-shifting transformer is shown in Fig. 5.2-2, where the primary winding remains the same as that in the Y/Z-1 transformer while the secondary delta-connected coils are connected in a reverse order. Following the same procedure presented earlier, the transformer turns ratio can be found from

$$\frac{N_3}{N_2 + N_3} = \frac{\sin(30° - |\delta|)}{\sin(30° + |\delta|)}$$

$$\frac{N_1}{N_2 + N_3} = \frac{1}{2\sin(30° + |\delta|)} \cdot \frac{V_{AB}}{V_{ab}} \qquad -30° \leq \delta \leq 0° \tag{5.2-9}$$

The phase angle δ has a negative value for the Y/Z-2 transformer, indicating that V_{ab} lags V_{AB} by $|\delta|$ as shown in Fig. 5.2-2b.

Table 5.2-1 gives the typical value of δ and turns ratio of the Y/Z transformers for use in multipulse rectifiers. The voltage ratio V_{AB}/V_{ab} is normally equal to 2, 3, and 4 for the 12-, 18-, and 24-pulse rectifiers, respectively.

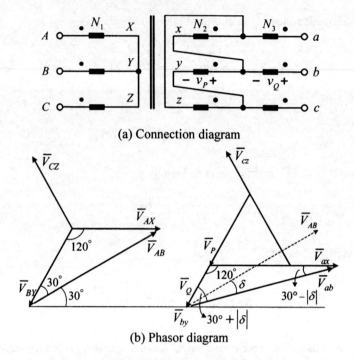

(a) Connection diagram

(b) Phasor diagram

Figure 5.2-2 Y/Z-2 phase-shifting transformer.

Table 5.2-1 Turns Ratio for Y/Z Transformers

δ ($\angle \overline{V}_{ab} - \angle \overline{V}_{AB}$)		$\dfrac{N_3}{N_2 + N_3}$	$\dfrac{N_1}{N_2 + N_3}$	Applications
Y/Z-1	Y/Z-2			
0°	0°	1.0	$1.0 \dfrac{V_{AB}}{V_{ab}}$	12-, 18-, and 24-pulse rectifiers
15°	−15°	0.366	$0.707 \dfrac{V_{AB}}{V_{ab}}$	24-pulse rectifiers
20°	−20°	0.227	$0.653 \dfrac{V_{AB}}{V_{ab}}$	18-pulse rectifiers
30°	−30°	0	$0.577 \dfrac{V_{AB}}{V_{ab}}$	12- and 24-pulse rectifiers

$\dfrac{V_{AB}}{V_{ab}} = 2, 3,$ and 4 for 12-, 18-, and 24-pulse rectifiers, respectively.

5.3 Δ/Z TRANSFORMERS

Figure 5.3-1 shows two typical configurations for Δ/Z phase shifting transformers, where the primary winding is connected in delta and the secondary winding is zigzag-connected. The phasor diagram for the Δ/Z – 1 transformer is given in Fig. 5.3-1c, in which the secondary voltage V_{ab} lags the primary voltage V_{AB} by $|\delta|$. The turns ratio of the Δ/Z – 1 transformer is given by

$$\frac{N_3}{N_2+N_3} = \frac{V_Q}{V_{ax}} = \frac{\sin(|\delta|)}{\sin(60° - |\delta|)}$$

$$\frac{N_1}{N_2+N_3} = \frac{V_{AX}}{V_{ax}} = \frac{\sqrt{3}}{2\sin(60° - |\delta|)} \frac{V_{AB}}{V_{ab}}$$

$$-30° \leq \delta \leq 0° \qquad (5.3\text{-}1)$$

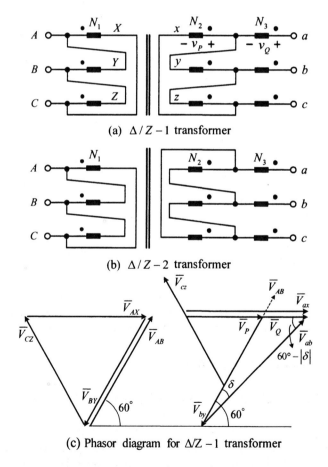

(a) Δ/Z – 1 transformer

(b) Δ/Z – 2 transformer

(c) Phasor diagram for Δ/Z – 1 transformer

Figure 5.3-1 Δ/Z phase-shifting transformers.

Table 5.3-1 Turns Ratio for Δ/Z Transformers

Zigzag Transformer	δ ($\angle \overline{V}_{ab} - \angle \overline{V}_{AB}$)	$\dfrac{N_3}{N_2 + N_3}$	$\dfrac{N_1}{N_2 + N_3}$	Applications
Δ/Z-1	0°	0	$1.0 \dfrac{V_{AB}}{V_{ab}}$	12-, 18-, and 24-pulse rectifiers
	−15°	0.366	$1.225 \dfrac{V_{AB}}{V_{ab}}$	24-pulse rectifiers
	−20°	0.532	$1.347 \dfrac{V_{AB}}{V_{ab}}$	18-pulse rectifiers
	−30°	1.0	$1.732 \dfrac{V_{AB}}{V_{ab}}$	12- and 24-pulse rectifiers
Δ/Z-2	−40°	0.532	$1.347 \dfrac{V_{AB}}{V_{ab}}$	18-pulse rectifiers
	−45°	0.366	$1.225 \dfrac{V_{AB}}{V_{ab}}$	24-pulse rectifiers
	−60°	0	$1.0 \dfrac{V_{AB}}{V_{ab}}$	18-pulse rectifiers

$\dfrac{V_{AB}}{V_{ab}} = 2, 3,$ and 4 for 12-, 18-, and 24-pulse rectifiers, respectively.

Table 5.3-1 illustrates the relationship between the phase shifting angle δ and the turns ratio of the Δ/Z transformers for multipulse rectifiers. The phase angle δ varies from 0° to −30° for the Δ/Z − 1 transformer and from −30° to −60° for the Δ/Z − 2 transformer.

Figure 5.3-2 shows a few examples of the phase-shifting transformers for use in multipulse diode/SCR rectifiers. The transformer for the 12-pulse rectifiers has two secondary windings with a 30° phase shift between them. The 18-pulse rectifiers require a transformer with three secondary windings having a 20° phase displacement among each other. The transformer used in the 24-pulse rectifiers has four secondary windings with a 15° phase shift between any two adjacent windings.

5.4 HARMONIC CURRENT CANCELLATION

5.4.1 Phase Displacement of Harmonic Currents

The main purpose of this section is to investigate the phase displacement of harmonic currents when they are referred from the secondary to the primary side of a phase shifting transformer. It is the phase displacement that makes it possible to cancel certain harmonic currents generated by a three-phase nonlinear load.

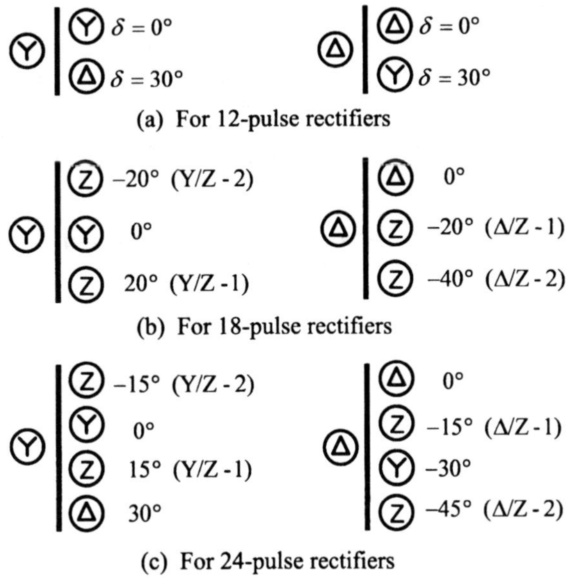

5.4 Harmonic Current Cancellation

(a) For 12-pulse rectifiers

Ⓨ | Ⓨ δ = 0°
  | Ⓐ δ = 30°

Ⓐ | Ⓐ δ = 0°
  | Ⓨ δ = 30°

(b) For 18-pulse rectifiers

Ⓨ | Ⓩ −20° (Y/Z - 2)
  | Ⓨ 0°
  | Ⓩ 20° (Y/Z -1)

Ⓐ | Ⓐ 0°
  | Ⓩ −20° (Δ/Z - 1)
  | Ⓩ −40° (Δ/Z - 2)

(c) For 24-pulse rectifiers

Ⓨ | Ⓩ −15° (Y/Z - 2)
  | Ⓨ 0°
  | Ⓩ 15° (Y/Z -1)
  | Ⓐ 30°

Ⓐ | Ⓐ 0°
  | Ⓩ −15° (Δ/Z -1)
  | Ⓨ −30°
  | Ⓩ −45° (Δ/Z - 2)

Figure 5.3-2 Examples of phase-shifting transformers for multipulse rectifiers.

Figure 5.4-1 shows a Δ/Y transformer feeding a nonlinear load. Assume that the voltage ratio V_{AB}/V_{ab} of the transformer is unity with a turns ratio of $N_1/N_2 = \sqrt{3}$. The transformer has a phase angle of $\delta = \angle \overline{V}_{ab} - \angle \overline{V}_{AB} = -30°$. For a three-phase balanced system, the line currents of the nonlinear load can be expressed as

$$i_a = \sum_{n=1,5,7,11,\ldots}^{\infty} \hat{I}_n \sin(n\omega t)$$

$$i_b = \sum_{n=1,5,7,11,\ldots}^{\infty} \hat{I}_n \sin(n(\omega t - 120°)) \quad (5.4\text{-}1)$$

$$i_c = \sum_{n=1,5,7,11,\ldots}^{\infty} \hat{I}_n \sin(n(\omega t - 240°))$$

where $\hat{I}_n$ is the peak value of the nth order harmonic current. When i_a and i_b are referred to the primary side, the referred currents i'_{ap} and i'_{bp} in the primary winding can be described by

$$i'_{ap} = i_a \frac{N_2}{N_1} = \frac{1}{\sqrt{3}}(\hat{I}_1 \sin(\omega t) + \hat{I}_5 \sin(5\omega t) + \hat{I}_7 \sin(7\omega t) + \hat{I}_{11} \sin(11\omega t) + \cdots)$$

$$i'_{bp} = i_b \frac{N_2}{N_1} = \frac{1}{\sqrt{3}}(\hat{I}_1 \sin(\omega t - 120°) + \hat{I}_5 \sin(5\omega t - 240°) + \hat{I}_7 \sin(7\omega t - 120°)$$

$$+ \hat{I}_{11} \sin(11\omega t - 240°) + \cdots) \quad (5.4\text{-}2)$$

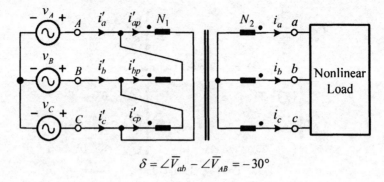

$$\delta = \angle \overline{V}_{ab} - \angle \overline{V}_{AB} = -30°$$

Figure 5.4-1 Investigation of harmonic currents in the primary and secondary windings.

from which the primary line current can be found from

$$i'_a = i'_{ap} - i'_{bp} = \hat{I}_1 \sin(\omega t + 30°) + \hat{I}_5 \sin(5\omega t - 30°) + \hat{I}_7 \sin(7\omega t + 30°)$$

$$+ \hat{I}_{11} \sin(11\omega t - 30°) + \cdots \quad (5.4\text{-}3)$$

$$= \sum_{n=1,7,13,\ldots}^{\infty} \hat{I}_n \sin(n\omega t - \delta) + \sum_{n=5,11,17,\ldots}^{\infty} \hat{I}_n \sin(n\omega t + \delta)$$

The first Σ on the right-hand side of (5.4-3) includes all the harmonic currents of **positive sequence** ($n = 1, 7, 13, \ldots$), while the second Σ represents all the **negative sequence** harmonics ($n = 5, 11, 17, \ldots$).

Comparing the primary line current i'_a in (5.4-3) with the second line current i_a in (5.4-1), we have

$$\angle i'_{an} = \angle i_{an} - \delta \quad \text{for } n = 1, 7, 13, 19, \ldots \text{ (positive sequence harmonics)}$$
$$\angle i'_{an} = \angle i_{an} + \delta \quad \text{for } n = 5, 11, 17, 23, \ldots \text{ (negative sequence harmonics)} \quad (5.4\text{-}4)$$

where $\angle i'_{an}$ and $\angle i_{an}$ are the phase angles of nth-order harmonic currents i'_{an} and i_{an}, respectively. Equation (5.4-4) describes the relationship between the phase angles of the harmonic currents when referred from the secondary to the primary of the phase shifting transformer. It can be proven that Eq. (5.4-4) is valid for any δ values.

5.4.2 Harmonic Cancellation

To illustrate how the harmonic currents are canceled by a phase shifting transformer, let's examine a 12-pulse rectifier shown in Fig. 5.4-2. The phase-shifting angle δ of the wye- and delta-connected secondary windings is 0° and 30°, respec-

5.4 Harmonic Current Cancellation

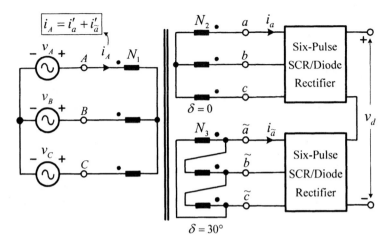

Figure 5.4-2 An example of harmonic current cancellation.

tively. The voltage ratio is $V_{AB}/V_{ab} = V_{AB}/V_{\tilde{a}\tilde{b}} = 2$. The line currents in the secondary windings can be expressed as

$$i_a = \sum_{n=1,5,7,11,13,\ldots}^{\infty} \hat{I}_n \sin(n\omega t) \cdots$$

$$i_{\tilde{a}} = \sum_{n=1,5,7,11,13,\ldots}^{\infty} \hat{I}_n \sin(n(\omega t + \delta))$$

(5.4-5)

When i_a is referred to the primary side, the phase angle of all the harmonic currents remains unchanged due to the Y/Y connection. The referred current i'_a is then given by

$$i'_a = \tfrac{1}{2}\left(\hat{I}_1 \sin(\omega t) + \hat{I}_5 \sin(5\omega t) + \hat{I}_7 \sin(7\omega t) + \hat{I}_{11} \sin(11\omega t) + \hat{I}_{13} \sin(13\omega t) + \cdots\right)$$
(5.4-6)

To transfer $i_{\tilde{a}}$ to the primary side, we can make use of (5.4-4), from which

$$i'_{\tilde{a}} = \frac{1}{2}\left(\sum_{n=1,7,13,\ldots}^{\infty} \hat{I}_n \sin(n(\omega t + \delta) - \delta) + \sum_{n=5,11,17,\ldots}^{\infty} \hat{I}_n \sin(n(\omega t + \delta) + \delta)\right)$$

$$= \tfrac{1}{2}(\hat{I}_1 \sin(\omega t) - \hat{I}_5 \sin(5\omega t) - \hat{I}_7 \sin(7\omega t) + \hat{I}_{11} \sin(11\omega t) + \hat{I}_{13} \sin(13\omega t) - \cdots)$$

for $\delta = 30°$ (5.4-7)

The primary line current i_A can then be found from

$$i_A = i'_a + i'_{\tilde{a}} = \hat{I}_1 \sin \omega t + \hat{I}_{11} \sin 11\omega t + \hat{I}_{13} \sin 13\omega t + \hat{I}_{23} \sin 23\omega t + \cdots \quad (5.4-8)$$

where the 5th, 7th, 17th, and 19th harmonic currents in i_a and i'_a are 180° out of phase, and therefore canceled.

5.5 SUMMARY

To reduce the line current THD in high-power rectifiers, multipulse diode/SCR rectifiers powered by phase-shifting transformers are often employed. In this chapter, the typical configurations of the phase-shifting transformers for 12-, 18-, and 24-pulse rectifiers are presented. The structure and phasor diagrams of the transformers are discussed. To assist the transformer design, the relationship between the required phase-shifting angle and transformer turns ratio is tabulated. The principle of harmonic current cancellation by the phase-shifting transformers is also demonstrated.

Part Three

Multilevel Voltage Source Converters

Chapter 6

Two-Level Voltage Source Inverter

6.1 INTRODUCTION

The primary function of a voltage source inverter (VSI) is to convert a fixed dc voltage to a three-phase ac voltage with variable magnitude and frequency. A simplified circuit diagram for a two-level voltage source inverter for high-power medium-voltage applications is shown in Fig. 6.1-1. The inverter is composed of six group of active switches, $S_1 \sim S_6$, with a free-wheeling diode in parallel with each switch. Depending on the dc operating voltage of the inverter, each switch group consists of two or more IGBT or GCT switching devices connected in series.

This chapter focuses on **pulse width modulation** (PWM) schemes for the high-power two-level inverter, where the device switching frequency is normally below 1 kHz. A carrier-based sinusoidal PWM (SPWM) scheme is reviewed, followed by a detailed analysis on **space vector modulation** (SVM) algorithms. The conventional SVM scheme usually generates both even- and odd-order harmonics voltages. The mechanism of even-order harmonic generation is analyzed, and a modified SVM scheme for even-order harmonic elimination is presented.

6.2 SINUSOIDAL PWM

6.2.1 Modulation Scheme

The principle of the sinusoidal PWM scheme for the two-level inverter is illustrated in Fig. 6.2-1, where v_{mA}, v_{mB}, and v_{mC} are the three-phase sinusoidal **modulating waves** and v_{cr} is the triangular **carrier wave**. The fundamental-frequency component in the inverter output voltage can be controlled by **amplitude modulation index**

$$m_a = \frac{\hat{V}_m}{\hat{V}_{cr}} \qquad (6.2\text{-}1)$$

Chapter 6 Two-Level Voltage Source Inverter

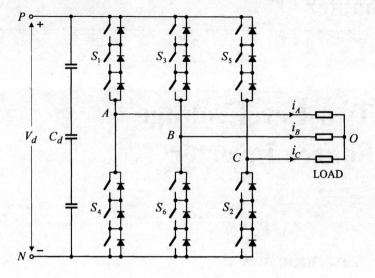

Figure 6.1-1 Simplified two-level inverter for high-power applications.

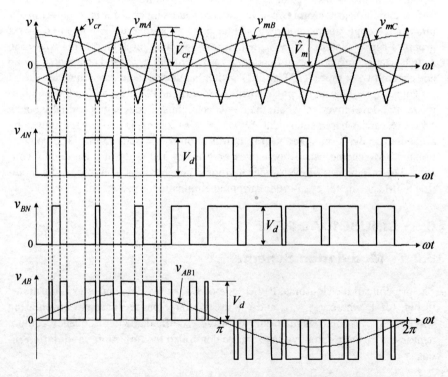

Figure 6.2-1 Sinusoidal pulse-width modulation (SPWM).

where $\hat{V}_m$ and $\hat{V}_{cr}$ are the peak values of the modulating and carrier waves, respectively. The amplitude modulation index m_a is usually adjusted by varying $\hat{V}_m$ while keeping $\hat{V}_{cr}$ fixed. The **frequency modulation index** is defined by

$$m_f = \frac{f_{cr}}{f_m} \qquad (6.2\text{-}2)$$

where f_m and f_{cr} are the frequencies of the modulating and carrier waves, respectively.

The operation of switches S_1 to S_6 is determined by comparing the modulating waves with the carrier wave. When $v_{mA} \geq v_{cr}$, the upper switch S_1 in inverter leg A is turned on. The lower switch S_4 operates in a complementary manner and thus is switched off. The resultant **inverter terminal voltage** v_{AN}, which is the voltage at the phase A terminal with respect to the negative dc bus N, is equal to the dc voltage V_d. When $v_{mA} < v_{cr}$, S_4 is on and S_1 is off, leading to $v_{AN} = 0$ as shown in Fig. 6.2-1. Since the waveform of v_{AN} has only two levels, V_d and 0, the inverter is known as a **two-level inverter**. It should be noted that to avoid possible short circuit during switching transients of the upper and lower devices in an inverter leg, a **blanking time** should be implemented, during which both switches are turned off.

The inverter line-to-line voltage v_{AB} can be determined by $v_{AB} = v_{AN} - v_{BN}$. The waveform of its fundamental-frequency component v_{AB1} is also given in the figure. The magnitude and frequency of v_{AB1} can be independently controlled by m_a and f_m, respectively.

The **switching frequency** of the active switches in the two-level inverter can be found from $f_{sw} = f_{cr} = f_m \times m_f$. For instance, v_{AN} in Fig. 6.2-1 contains nine pulses per cycle of the fundamental frequency. Each pulse is produced by turning S_1 on and off once. With the fundamental frequency of 60 Hz, the resultant switching frequency for S_1 is $f_{sw} = 60 \times 9 = 540$ Hz, which is also the carrier frequency f_{cr}. It is worth noting that the device switching frequency may not always be equal to the carrier frequency in multilevel inverters. This issue will be addressed in the later chapters.

When the carrier wave is synchronized with the modulating wave (m_f is an integer), the modulation scheme is known as **synchronous PWM**, in contrast to **asynchronous PWM** whose carrier frequency f_{cr} is usually fixed and independent of f_m. The asynchronous PWM features a fixed switching frequency and easy implementation with analog circuits. However, it may generate **noncharacteristic harmonics**, whose frequency is not a multiple of the fundamental frequency. The synchronous PWM scheme is more suitable for implementation with a digital processor.

6.2.2 Harmonic Content

Figure 6.2-2 shows a set of simulated waveforms for the two-level inverter, where v_{AB} is the inverter line-to-line voltage, v_{AO} is the **load phase voltage** and i_A is the load current. The inverter operates under the condition of $m_a = 0.8$, $m_f = 15$, $f_m = 60$ Hz, and $f_{sw} = 900$ Hz with a rated three-phase inductive load. The load power factor is 0.9 per phase. We can observe the following:

- All the harmonics in v_{AB} with the order lower than $(m_f - 2)$ are eliminated.
- The harmonics are centered around m_f and its multiples such as $2m_f$ and $3m_f$.

The above statements are valid for $m_f \geq 9$ provided that m_f is a multiple of 3 [1].

The waveform of the load current i_A is close to sinusoidal with a THD of 7.73%. The low amount of harmonic distortion is due to the elimination of low-order harmonics by the modulation scheme and the filtering effect of the load inductance.

Figure 6.2-3 shows the harmonic content of the inverter line-to-line voltage v_{AB} normalized to its dc voltage V_d as a function of m_a, where V_{ABn} is the nth-order harmonic voltage (rms). The fundamental-frequency component V_{AB1} increases linearly with m_a, whose maximum value can be found from

$$V_{AB1,\max} = 0.612 V_d \quad \text{for } m_a = 1 \quad (6.2\text{-}3)$$

The THD curve for v_{AB} is also given in the figure.

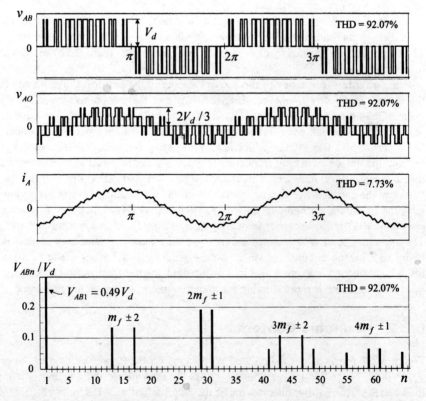

Figure 6.2-2 Simulated waveforms for the two-level inverter operating at $m_a = 0.8$, $m_f = 15$, $f_m = 60$ Hz, and $f_{sw} = 900$ Hz.

The injected third harmonic component v_{m3} will not increase the harmonic distortion for v_{AB}. Although it appears in each of the inverter terminal voltages v_{AN}, v_{BN} and v_{CN}, the third-order harmonic voltage does not exist in the line-to-line voltage v_{AB}. This is because the line-to-line voltage is given by $v_{AB} = v_{AN} - v_{BN}$, where the third-order harmonics in v_{AN} and v_{BN} are of zero sequence with the same magnitude and phase displacement and thus cancel each other.

6.3 SPACE VECTOR MODULATION

Space vector modulation (SVM) is one of the preferred real-time modulation techniques and is widely used for digital control of voltage source inverters [3, 4]. This section presents the principle and implementation of the space vector modulation for the two-level inverter.

6.3.1 Switching States

The operating status of the switches in the two-level inverter in Fig. 6.1-1 can be represented by **switching states**. As indicated in Table 6.3-1, switching state 'P' denotes that the upper switch in an inverter leg is on and the inverter terminal voltage (v_{AN}, v_{BN}, or v_{CN}) is positive ($+V_d$) while 'O' indicates that the inverter terminal voltage is zero due to the conduction of the lower switch.

There are eight possible combinations of switching states in the two-level inverter as listed in Table 6.3-2. The switching state [POO], for example, corresponds to the conduction of S_1, S_6, and S_2 in the inverter legs A, B, and C, respectively. Among the eight switching states, [PPP] and [OOO] are **zero states** and the others are **active states**.

6.3.2 Space Vectors

The active and zero switching states can be represented by active and zero **space vectors**, respectively. A typical space vector diagram for the two-level inverter is shown in Fig. 6.3-1, where the six **active vectors** $\vec{V}_1$ to $\vec{V}_6$ form a regular hexagon with six equal sectors (I to VI). The **zero vector** $\vec{V}_0$ lies on the center of the hexagon.

Table 6.3-1 Definition of Switching States

Switching State	Leg A			Leg B			Leg C		
	S_1	S_4	v_{AN}	S_3	S_6	v_{BN}	S_5	S_2	v_{CN}
P	On	Off	V_d	On	Off	V_d	On	Off	V_d
O	Off	On	0	Off	On	0	Off	On	0

102   Chapter 6   Two-Level Voltage Source Inverter

Table 6.3-2   Space Vectors, Switching States, and On-State Switches

Space Vector		Switching State (Three Phases)	On-State Switch	Vector Definition
Zero Vector	$\vec{V}_0$	[PPP]	S_1, S_3, S_5	$\vec{V}_0 = 0$
		[OOO]	S_4, S_6, S_2	
Active Vector	$\vec{V}_1$	[POO]	S_1, S_6, S_2	$\vec{V}_1 = \dfrac{2}{3} V_d e^{j0}$
	$\vec{V}_2$	[PPO]	S_1, S_3, S_2	$\vec{V}_2 = \dfrac{2}{3} V_d e^{j\frac{\pi}{3}}$
	$\vec{V}_3$	[OPO]	S_4, S_3, S_2	$\vec{V}_3 = \dfrac{2}{3} V_d e^{j\frac{2\pi}{3}}$
	$\vec{V}_4$	[OPP]	S_4, S_3, S_5	$\vec{V}_4 = \dfrac{2}{3} V_d e^{j\frac{3\pi}{3}}$
	$\vec{V}_5$	[OOP]	S_4, S_6, S_5	$\vec{V}_5 = \dfrac{2}{3} V_d e^{j\frac{4\pi}{3}}$
	$\vec{V}_6$	[POP]	S_1, S_6, S_5	$\vec{V}_6 = \dfrac{2}{3} V_d e^{j\frac{5\pi}{3}}$

To derive the relationship between the space vectors and switching states, refer to the two-level inverter in Fig. 6.1-1. Assuming that the operation of the inverter is three-phase balanced, we have

$$v_{AO}(t) + v_{BO}(t) + v_{CO}(t) = 0 \qquad (6.3\text{-}1)$$

where v_{AO}, v_{BO}, and v_{CO} are the instantaneous load phase voltages. From mathematical point of view, one of the phase voltages is redundant since given any two phase

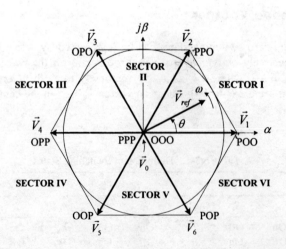

Figure 6.3-1   Space vector diagram for the two-level inverter.

6.3 Space Vector Modulation

voltages, the third one can be readily calculated. Therefore, it is possible to transform the three-phase variables to equivalent two-phase variables [5]:

$$\begin{bmatrix} v_\alpha(t) \\ v_\beta(t) \end{bmatrix} = \frac{2}{3} \begin{bmatrix} 1 & -\frac{1}{2} & -\frac{1}{2} \\ 0 & \frac{\sqrt{3}}{2} & -\frac{\sqrt{3}}{2} \end{bmatrix} \begin{bmatrix} v_{AO}(t) \\ v_{BO}(t) \\ v_{CO}(t) \end{bmatrix} \quad (6.3\text{-}2)$$

The coefficient 2/3 is somewhat arbitrarily chosen. The commonly used value is 2/3 or $\sqrt{2/3}$. The main advantage of using 2/3 is that the magnitude of the two-phase voltages will be equal to that of the three-phase voltages after the transformation. A space vector can be generally expressed in terms of the two-phase voltages in the α–β plane

$$\vec{V}(t) = v_\alpha(t) + jv_\beta(t) \quad (6.3\text{-}3)$$

Substituting (6.3-2) into (6.3-3), we have

$$\vec{V}(t) = \frac{2}{3}[v_{AO}(t)e^{j0} + v_{BO}(t)e^{j2\pi/3} + v_{CO}(t)e^{j4\pi/3}] \quad (6.3\text{-}4)$$

where $e^{jx} = \cos x + j\sin x$ and $x = 0, 2\pi/3$ or $4\pi/3$. For active switching state [POO], the generated load phase voltages are

$$v_{AO}(t) = \frac{2}{3}V_d, \quad v_{BO}(t) = -\frac{1}{3}V_d, \quad \text{and} \quad v_{CO}(t) = -\frac{1}{3}V_d \quad (6.3\text{-}5)$$

The corresponding space vector, denoted as $\vec{V}_1$, can be obtained by substituting (6.3-5) into (6.3-4):

$$\vec{V}_1 = \frac{2}{3} V_d e^{j0} \quad (6.3\text{-}6)$$

Following the same procedure, all six active vectors can be derived

$$\vec{V}_k = \frac{2}{3} V_d e^{j(k-1)\frac{\pi}{3}}, \quad k = 1, 2, \ldots, 6 \quad (6.3\text{-}7)$$

The zero vector $\vec{V}_0$ has two switching states [PPP] and [OOO], one of which seems redundant. As will be seen later, the **redundant switching state** can be utilized to minimize the switching frequency of the inverter or perform other useful functions. The relationship between the space vectors and their corresponding switching states is given in Table 6.3-2.

Note that the zero and active vectors do not move in space, and thus they are referred to as **stationary vectors**. On the contrary, the **reference vector** $\vec{V}_{ref}$ in Fig. 6.3-1 rotates in space at an angular velocity

$$\omega = 2\pi f_1 \qquad (6.3\text{-}8)$$

where f_1 is the fundamental frequency of the inverter output voltage. The angular displacement between $\vec{V}_{ref}$ and the α-axis of the α–β plane can be obtained by

$$\theta(t) = \int_0^t \omega(t)dt + \theta(0) \qquad (6.3\text{-}9)$$

For a given magnitude (length) and position, $\vec{V}_{ref}$ can be synthesized by three nearby stationary vectors, based on which the switching states of the inverter can be selected and gate signals for the active switches can be generated. When $\vec{V}_{ref}$ passes through sectors one by one, different sets of switches will be turned on or off. As a result, when $\vec{V}_{ref}$ rotates one revolution in space, the inverter output voltage varies one cycle over time. The inverter output frequency corresponds to the rotating speed of $\vec{V}_{ref}$, while its output voltage can be adjusted by the magnitude of $\vec{V}_{ref}$.

6.3.3 Dwell Time Calculation

As mentioned earlier, the reference $\vec{V}_{ref}$ can be synthesized by three stationary vectors. The dwell time for the stationary vectors essentially represents the duty-cycle time (on-state or off-state time) of the chosen switches during a **sampling period** T_s of the modulation scheme. The dwell time calculation is based on '**volt-second balancing**' principle, that is, the product of the reference voltage $\vec{V}_{ref}$ and sampling period T_s equals the sum of the voltage multiplied by the time interval of chosen space vectors.

Assuming that the sampling period T_s is sufficiently small, the reference vector $\vec{V}_{ref}$ can be considered constant during T_s. Under this assumption, $\vec{V}_{ref}$ can be approximated by two adjacent active vectors and one zero vector. For example, when $\vec{V}_{ref}$

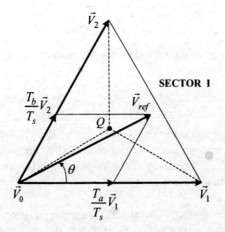

Figure 6.3-2 $\vec{V}_{ref}$ synthesized by $\vec{V}_1, \vec{V}_2$ and $\vec{V}_0$.

falls into sector I as shown in Fig. 6.3-2, it can be synthesized by $\vec{V}_1$, $\vec{V}_2$, and $\vec{V}_0$. The volt-second balancing equation is

$$\vec{V}_{ref}T_s = \vec{V}_1 T_a + \vec{V}_2 T_b + \vec{V}_0 T_0$$
$$T_s = T_a + T_b + T_0 \tag{6.3-10}$$

where T_a, T_b, and T_0 are the **dwell times** for the vectors $\vec{V}_1$, $\vec{V}_2$ and $\vec{V}_0$, respectively. The space vectors in (6.3-10) can be expressed as

$$\vec{V}_{ref} = \vec{V}_{ref}e^{j\theta}, \quad \vec{V}_1 = \frac{2}{3}V_d, \quad \vec{V}_2 = \frac{2}{3}V_d e^{j\frac{\pi}{3}}, \quad \text{and} \quad \vec{V}_0 = 0 \tag{6.3-11}$$

Substituting (6.3-11) into (6.3-10) and then splitting the resultant equation into the real (β-axis) and imaginary (β-axis) components in the α–β plane, we have

$$\text{Re:} \quad V_{ref}(\cos\theta)T_s = \frac{2}{3}V_d T_a + \frac{1}{3}V_d T_b$$
$$\text{Im:} \quad V_{ref}(\sin\theta)T_s = \frac{1}{\sqrt{3}}V_d T_b \tag{6.3-12}$$

Solving (6.3-12) together with $T_s = T_a + T_b + T_0$ yields

$$T_a = \frac{\sqrt{3}T_s V_{ref}}{V_d}\sin\left(\frac{\pi}{3} - \theta\right)$$
$$T_b = \frac{\sqrt{3}T_s V_{ref}}{V_d}\sin\theta \quad \text{for } 0 \le \theta < \pi/3 \tag{6.3-13}$$
$$T_0 = T_s - T_a - T_b$$

To visualize the relationship between the location of $\vec{V}_{ref}$ and the dwell times, let us examine some special cases. If $\vec{V}_{ref}$ lies exactly in the middle between $\vec{V}_1$ and $\vec{V}_2$ (i.e., $\theta = \pi/6$), the dwell time T_a for $\vec{V}_1$ will be equal to T_b for $\vec{V}_2$. When $\vec{V}_{ref}$ is closer to $\vec{V}_2$ than $\vec{V}_1$, T_b will be greater than T_a. If $\vec{V}_{ref}$ is coincident with $\vec{V}_2$, T_a will be zero. With the head of $\vec{V}_{ref}$ located right on the central point Q in figure 6.3-2, $T_a = T_b = T_0$. The relationship between the $\vec{V}_{ref}$ location and dwell times is summarized in Table 6.3-3.

Note that although Eq. (6.3-13) is derived when $\vec{V}_{ref}$ is in sector I, it can also be used when $\vec{V}_{ref}$ is in other sectors provided that a multiple of $\pi/3$ is subtracted from the actual angular displacement θ such that the modified angle θ' falls into the range between zero and $\pi/3$ for use in the equation, that is,

$$\theta' = \theta - (k-1)\pi/3 \quad \text{for } 0 \le \theta' < \pi/3 \tag{6.3-14}$$

where $k = 1, 2, \ldots, 6$ for sectors I, II, $\ldots$, VI, respectively. For example, when $\vec{V}_{ref}$ is in sector II, the calculated dwell times T_a, T_b, and T_0 based on (6.3-13) and (6.3-14) are for vectors $\vec{V}_2$, $\vec{V}_3$, and $\vec{V}_0$, respectively.

Table 6.3-3 $\vec{V}_{ref}$ Location and Dwell Times

$\vec{V}_{ref}$ Location:	$\theta = 0$	$0 < \theta < \dfrac{\pi}{6}$	$\theta = \dfrac{\pi}{6}$	$\dfrac{\pi}{6} < \theta < \dfrac{\pi}{3}$	$\theta = \dfrac{\pi}{3}$
Dwell Times:	$T_a > 0$ $T_b = 0$	$T_a > T_b$	$T_a = T_b$	$T_a < T_b$	$T_a = 0$ $T_b > 0$

6.3.4 Modulation Index

Equation (6.3-13) can be also expressed in terms of **modulation index** m_a

$$T_a = T_s m_a \sin\left(\frac{\pi}{3} - \theta\right)$$

$$T_b = T_s m_a \sin\theta \qquad (6.3\text{-}15)$$

$$T_0 = T_s - T_a - T_b$$

where

$$m_a = \frac{\sqrt{3}V_{ref}}{V_d} \qquad (6.3\text{-}16)$$

The maximum magnitude of the reference vector, $V_{ref,max}$, corresponds to the radius of the largest circle that can be inscribed within the hexagon shown in Fig. 6.3-1. Since the hexagon is formed by six active vectors having a length of $2V_d/3$, $V_{ref,max}$ can be found from

$$V_{ref,max} = \frac{2}{3}V_d \times \frac{\sqrt{3}}{2} = \frac{V_d}{\sqrt{3}} \qquad (6.3\text{-}17)$$

Substituting (6.3-17) into (6.3-16) gives the maximum modulation index:

$$m_{a,max} = 1$$

from which the modulation index for the SVM scheme is in the range of

$$0 \leq m_a \leq 1 \qquad (6.3\text{-}18)$$

The maximum fundamental line-to-line voltage (rms) produced by the SVM scheme can be calculated by

$$V_{max,SVM} = \sqrt{3}(V_{ref,max}/\sqrt{2}) = 0.707 V_d \qquad (6.3\text{-}19)$$

where $V_{ref,max}/\sqrt{2}$ is the maximum rms value of the fundamental phase voltage of the inverter.

With the inverter controlled by the SPWM scheme, the maximum fundamental line-to-line voltage is

$$V_{max,SPWM} = 0.612 V_d \tag{6.3-20}$$

from which

$$\frac{V_{max,SVM}}{V_{max,SPWM}} = 1.155 \tag{6.3-21}$$

Equation (6.3-21) indicates that for a given dc bus voltage the maximum inverter line-to-line voltage generated by the SVM scheme is 15.5% higher than that by the SPWM scheme. However, the use of third harmonic injection SPWM scheme can also boost the inverter output voltage by 15.5%. Therefore, the two schemes have essentially the same **dc bus voltage utilization**.

6.3.5 Switching Sequence

With the space vectors selected and their dwell times calculated, the next step is to arrange switching sequence. In general, the switching sequence design for a given $\vec{V}_{ref}$ is not unique, but it should satisfy the following two requirements for the minimization of the device switching frequency:

(a) The transition from one switching state to the next involves only two switches in the same inverter leg, one being switched on and the other switched off.
(b) The transition for $\vec{V}_{ref}$ moving from one sector in the space vector diagram to the next requires no or minimum number of switchings.

Figure 6.3-3 shows a typical **seven-segment** switching sequence and inverter output voltage waveforms for $\vec{V}_{ref}$ in sector I, where $\vec{V}_{ref}$ is synthesized by $\vec{V}_1$, $\vec{V}_2$ and $\vec{V}_0$. The sampling period T_s is divided into seven segments for the selected vectors. The following can be observed:

- The dwell times for the seven segments add up to the sampling period ($T_s = T_a + T_b + T_0$).
- Design requirement (a) is satisfied. For instance, the transition from [OOO] to [POO] is accomplished by turning S_1 on and S_4 off, which involves only two switches.
- The redundant switching sates for $\vec{V}_0$ are utilized to reduce the number of switchings per sampling period. For the $T_0/4$ segment in the center of the sampling period, the switching state [PPP] is selected, whereas for the $T_0/4$ segments on both sides, the state [OOO] is used.
- Each of the switches in the inverter turns on and off once per sampling period. The **switching frequency** f_{sw} of the devices is thus equal to the **sampling frequency** f_{sp}, that is, $f_{sw} = f_{sp} = 1/T_s$.

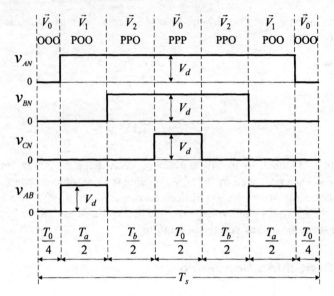

Figure 6.3-3 Seven-segment switching sequence for $\vec{V}_{ref}$ in sector I.

Let us now examine a case given in Fig. 6.3-4, where the vectors $\vec{V}_1$ and $\vec{V}_2$ in Fig. 6.3-3 are swapped. Some switching state transitions, such as the transition from [OOO] to [PPO], are accomplished by turning on and off four switches in two inverter legs simultaneously. As a consequence, the total number of switchings during the sampling period increases from six in the previous case to ten. Obviously, this switching sequence does not satisfy the design requirement and thus should not be adopted.

It is interesting to note that the waveforms of v_{AB} in Figs. 6.3-3 and 6.3-4 produced by two different switching sequences seem different, but they are essentially the same. If these two waveforms are drawn for two or more consecutive sampling periods, we will notice that they are identical except for a small time delay ($T_s/2$). Since T_s is much shorter than the period of the inverter fundamental frequency, the effect caused by the time delay is negligible.

Table 6.3-4 gives the seven-segment switching sequences for $\vec{V}_{ref}$ residing in all six sectors. Note that all the switching sequences start and end with switching state [OOO], which indicates that the transition for $\vec{V}_{ref}$ moving from one sector to the next does not require any switchings. The switching sequence design requirement (b) is satisfied.

6.3.6 Spectrum Analysis

The simulated waveforms for the inverter output voltages and load current are shown in Fig. 6.3-5. The inverter operates under the condition of $f_1 = 60$ Hz, $T_s = 1/720$ s, $f_{sw} = 720$ Hz, and $m_a = 0.8$ with a rated three-phase inductive load. The load power factor is 0.9 per phase. It can be observed that the waveform of the in-

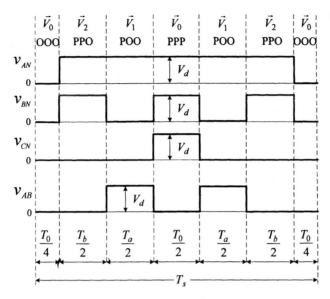

Figure 6.3-4 Undesirable seven-segment switching sequence.

verter line-to-line voltage v_{AB} is not **half-wave symmetrical**, that is, to $v_{AB}(\omega t) \neq -v_{AB}(\omega t + \pi)$. Therefore, it contains even-order harmonics, such as 2nd, 4th, 8th, and 10th, in addition to odd-order harmonics. The THD of v_{AB} and i_A is 80.2% and 8.37%, respectively.

Figure 6.3-6 shows waveforms measured from a laboratory two-level inverter operating under the same conditions as those given in Fig. 6.3-5. The top and bottom traces in Fig. 6.3-6a are the inverter line-to-line voltage v_{AB} and load phase

Table 6.3-4 Seven-Segment Switching Sequence

Sector	Switching Segment						
	1	2	3	4	5	6	7
I	$\vec{V}_0$ OOO	$\vec{V}_1$ POO	$\vec{V}_2$ PPO	$\vec{V}_0$ PPP	$\vec{V}_2$ PPO	$\vec{V}_1$ POO	$\vec{V}_0$ OOO
II	$\vec{V}_0$ OOO	$\vec{V}_3$ OPO	$\vec{V}_2$ PPO	$\vec{V}_0$ PPP	$\vec{V}_2$ PPO	$\vec{V}_3$ OPO	$\vec{V}_0$ OOO
III	$\vec{V}_0$ OOO	$\vec{V}_3$ OPO	$\vec{V}_4$ OPP	$\vec{V}_0$ PPP	$\vec{V}_4$ OPP	$\vec{V}_3$ OPO	$\vec{V}_0$ OOO
IV	$\vec{V}_0$ OOO	$\vec{V}_5$ OOP	$\vec{V}_4$ OPP	$\vec{V}_0$ PPP	$\vec{V}_4$ OPP	$\vec{V}_5$ OOP	$\vec{V}_0$ OOO
V	$\vec{V}_0$ OOO	$\vec{V}_5$ OOP	$\vec{V}_6$ POP	$\vec{V}_0$ PPP	$\vec{V}_6$ POP	$\vec{V}_5$ OOP	$\vec{V}_0$ OOO
VI	$\vec{V}_0$ OOO	$\vec{V}_1$ POO	$\vec{V}_6$ POP	$\vec{V}_0$ PPP	$\vec{V}_6$ POP	$\vec{V}_1$ POO	$\vec{V}_0$ OOO

110   Chapter 6   Two-Level Voltage Source Inverter

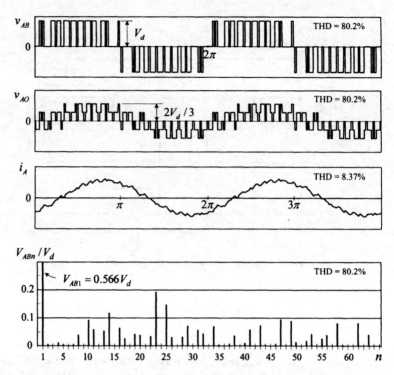

Figure 6.3-5 Inverter output waveforms produced by SVM scheme with $f_1 = 60$ Hz, $f_{sw} = 720$ Hz, and $m_a = 0.8$.

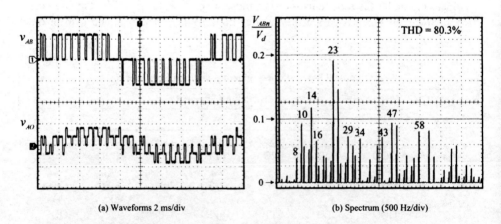

Figure 6.3-6 Measured inverter voltage waveforms and harmonic spectrum for the verification of simulated waveforms in Fig. 6.3-5.

voltage v_{AO}, and the spectrum of v_{AB} is given in Fig. 6.3-6b. The experimental results match with the simulation very well.

Figure 6.3-7 shows the harmonic content of v_{AB} for the inverter operating at $f_1 = $ 60 Hz and $f_{sw} = 720$ Hz. Although the low-order harmonics, such as 2nd, 4th, 5th, and 7th, are not eliminated, they have very low magnitudes. The maximum fundamental line-to-line voltage (rms) occurs at $m_a = 1$ and can be found from

$$V_{AB1,\max} = 0.707 V_d \qquad \text{for } m_a = 1 \qquad (6.3\text{-}22)$$

which is around 15.5% higher than that given in (6.2-3) for the SPWM scheme without using the third harmonic injection technique.

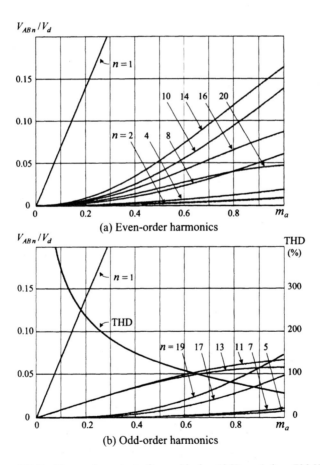

Figure 6.3-7 Harmonic content of v_{AB} with $f_1 = 60$ Hz and $f_{sw} = 720$ Hz.

6.3.7 Even-Order Harmonic Elimination

As indicated earlier, the line-to-line voltage waveform produced by the SVM inverter contains even-order harmonics. In the inverter-fed medium-voltage drives, these harmonics may not have a significant impact on the operation of the motor. However, when the two-level converter is used as a rectifier, its line current THD should comply with harmonic standards such as IEEE 519-1992. Since most standards have more stringent requirements on even-order harmonics than on odd-order ones, this section presents a modified SVM scheme with even-order harmonic elimination.

To investigate the mechanism of even-order harmonic generation, consider a case where the reference vector $\vec{V}_{ref}$ falls into sector IV. Based on the switching se-

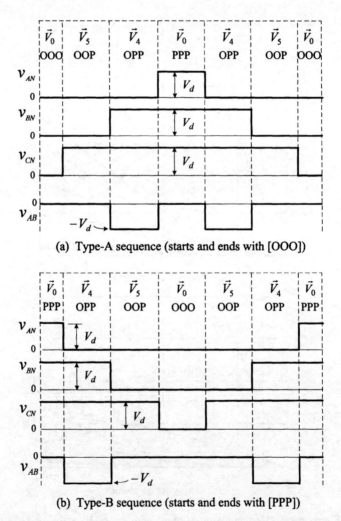

Figure 6.3-8 Two valid switching sequences for $\vec{V}_{ref}$ in sector IV.

quence given in Table 6.3-4, the waveform of inverter line-to-line voltage v_{AB} in a sampling period is illustrated in Fig. 6.3-8a. The waveform does not have a mirror image (not symmetrical about the horizontal axis) in comparison with that in Fig. 6.3-3, where $\vec{V}_{ref}$ is in a sector 180° apart from sector IV. This implies that the waveform generated by the SVM scheme is not half-wave symmetrical, leading to the generation of even-order harmonics.

Let's now consider type-B switching sequence shown in Fig. 6.3-8b, which is also a valid switching sequence that satisfies the design requirement (a) stated earlier. By comparing the waveform of v_{AB} with that in Fig. 6.3-3, it is clear that the use of this switching sequence would lead to $v_{AB}(\omega t) = -v_{AB}(\omega t + \pi)$. As a result, the waveform of v_{AB} would not contain any ever-order harmonics.

Examining the two switching sequences in Fig. 6.3-8, we can find out that the type-A sequence starts and ends with [OOO] while the type-B sequence commences and finishes with [PPP]. The waveforms of v_{AB} generated by both sequences seem different. However, they are essentially the same except for a small time delay ($T_s/2$), which can be clearly observed if these two waveforms are drawn for two or more consecutive sampling periods.

To make the three-phase line-to-line voltage half-wave symmetrical, type-A and type-B switching sequences can be alternatively used. In addition, each sector in the space vector diagram is divided into two regions as shown in Fig. 6.3-9. Type-A sequence is used in the nonshaded regions, while type-B sequence is employed in the shaded regions. The detailed switching sequence arrangements are given in Table 6.3-5.

It can be observed from the table that the transition for $\vec{V}_{ref}$ moving from region a to b causes additional switchings. This implies that the even-order harmonic elimination is achieved at the expense of an increase in switching frequency. The amount of switching frequency increase can be determined by

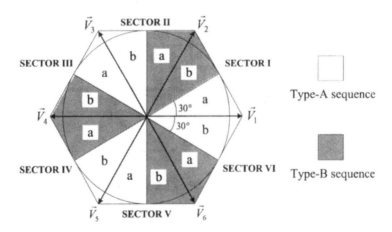

Figure 6.3-9 Alternative use of two switching sequences for even-order harmonic elimination.

Table 6.3-5 Switching Sequence of the Modified SVM for Even-Order Harmonic Elimination

Sector	Switching Sequence						
I-a	$\vec{V}_0$	$\vec{V}_1$	$\vec{V}_2$	$\vec{V}_0$	$\vec{V}_2$	$\vec{V}_1$	$\vec{V}_0$
	OOO	POO	PPO	PPP	PPO	POO	OOO
I-b	$\vec{V}_0$	$\vec{V}_2$	$\vec{V}_1$	$\vec{V}_0$	$\vec{V}_1$	$\vec{V}_2$	$\vec{V}_0$
	PPP	PPO	POO	OOO	POO	PPO	PPP
II-a	$\vec{V}_0$	$\vec{V}_2$	$\vec{V}_3$	$\vec{V}_0$	$\vec{V}_3$	$\vec{V}_2$	$\vec{V}_0$
	PPP	PPO	OPO	OOO	OPO	PPO	PPP
II-b	$\vec{V}_0$	$\vec{V}_3$	$\vec{V}_2$	$\vec{V}_0$	$\vec{V}_2$	$\vec{V}_3$	$\vec{V}_0$
	OOO	OPO	PPO	PPP	PPO	OPO	OOO
III-a	$\vec{V}_0$	$\vec{V}_3$	$\vec{V}_4$	$\vec{V}_0$	$\vec{V}_4$	$\vec{V}_3$	$\vec{V}_0$
	OOO	OPO	OPP	PPP	OPP	OPO	OOO
III-b	$\vec{V}_0$	$\vec{V}_4$	$\vec{V}_3$	$\vec{V}_0$	$\vec{V}_3$	$\vec{V}_4$	$\vec{V}_0$
	PPP	OPP	OPO	OOO	OPO	OPP	PPP
IV-a	$\vec{V}_0$	$\vec{V}_4$	$\vec{V}_5$	$\vec{V}_0$	$\vec{V}_5$	$\vec{V}_4$	$\vec{V}_0$
	PPP	OPP	OOP	OOO	OOP	OPP	PPP
IV-b	$\vec{V}_0$	$\vec{V}_5$	$\vec{V}_4$	$\vec{V}_0$	$\vec{V}_4$	$\vec{V}_5$	$\vec{V}_0$
	OOO	OOP	OPP	PPP	OPP	OOP	OOO
V-a	$\vec{V}_0$	$\vec{V}_5$	$\vec{V}_6$	$\vec{V}_0$	$\vec{V}_6$	$\vec{V}_5$	$\vec{V}_0$
	OOO	OOP	POP	PPP	POP	OOP	OOO
V-b	$\vec{V}_0$	$\vec{V}_6$	$\vec{V}_5$	$\vec{V}_0$	$\vec{V}_5$	$\vec{V}_6$	$\vec{V}_0$
	PPP	POP	OOP	OOO	OOP	POP	PPP
VI-a	$\vec{V}_0$	$\vec{V}_6$	$\vec{V}_1$	$\vec{V}_0$	$\vec{V}_1$	$\vec{V}_6$	$\vec{V}_0$
	PPP	POP	POO	OOO	POO	POP	PPP
VI-b	$\vec{V}_0$	$\vec{V}_1$	$\vec{V}_6$	$\vec{V}_0$	$\vec{V}_6$	$\vec{V}_1$	$\vec{V}_0$
	OOO	POO	POP	PPP	POP	POO	OOO

$$\Delta f_{sw} = 3f_1 \qquad (6.3\text{-}23)$$

where f_1 is the fundamental frequency of the inverter output voltage.

The inverter output waveforms measured from a laboratory two-level inverter with modified SVM scheme are shown in Fig. 6.3-10. The inverter operates under the condition of $f_1 = 60$ Hz, $T_s = 1/720$ s, and $m_a = 0.8$. The waveforms of the inverter line-to-line voltage v_{AB} and load phase voltage v_{AO} are of half-wave symme-

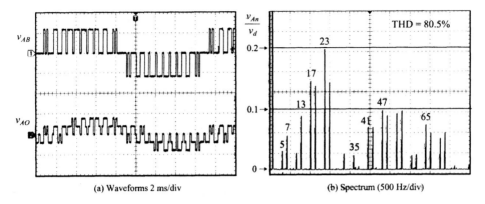

(a) Waveforms 2 ms/div   (b) Spectrum (500 Hz/div)

Figure 6.3-10 Measured waveforms produced by the modified SVM with even-order harmonic elimination ($f_1 = 60$ Hz, $T_s = 1/720$ s, and $m_a = 0.8$).

try, containing no even-order harmonics. Compared with the harmonic spectrum given in Fig. 6.3-6, the magnitude of the 5th and 7th harmonics in v_{AB} is increased while the THD essentially remains the same (refer to Appendix at the end of the book for details).

6.3.8 Discontinuous Space Vector Modulation

As pointed out earlier, the switching sequence design is not unique for a given set of stationary vectors and dwell times. Figure 6.3-11 shows two **five-segment** switching sequences and generated inverter terminal voltages for $\vec{V}_{ref}$ in sector I. For type-A sequence, the zero switching sate [OOO] is assigned for $\vec{V}_0$ while type-B sequence utilizes [PPP] for $\vec{V}_0$.

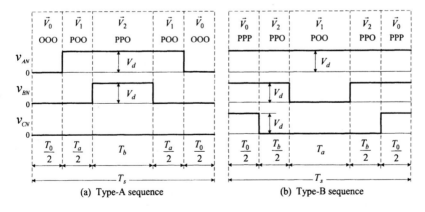

Figure 6.3-11 Five-segment switching sequence.

116   Chapter 6   Two-Level Voltage Source Inverter

In the five-segment sequence, one of the three inverter output terminals is clamped to either the positive or negative dc bus without any switchings during the sampling period T_s. Furthermore, the switching sequence can be arranged such that the switching in an inverter leg is continuously suppressed for a period of $2\pi/3$ per cycle of the fundamental frequency. For instance, the inverter terminal voltage v_{CN} can be clamped to the negative dc bus continuously in sectors I and II as shown in Table 6.3-6. Due to the switching discontinuity, the five-segment scheme is also known as **discontinuous space vector modulation** [4].

The use of type-A sequence alone will make the conduction angle of the lower switch in an inverter leg longer than that of the upper switch, causing unbalanced power and thermal distributions. The problem can be mitigated by swapping the two types of the switching sequences periodically. The switching frequency of the inverter will increase accordingly.

Figure 6.3-12 shows the simulated waveforms for v_{AB} and i_A when the inverter operates at f_1 = 60 Hz, f_{sw} = 600 Hz, T_s = 1/900 s, and m_a = 0.8 with a rated three-phase inductive load. The load power factor is 0.9 per phase. Since the gate signals for S_1, S_3 and S_5 are suppressed continuously for a period of $2\pi/3$ per cycle of the fundamental frequency, the switching frequency of the five-segment sequence is reduced by 1/3 compared with the seven-segment sequence with the same sampling period. The waveform of v_{AB} is not half-wave symmetrical, containing large amount of even-order harmonics. The THDs of v_{AB} and i_A are 91.8% and 12.1%, respectively, which are higher than those in the seven-segment sequence. This is mainly caused by the reduction of switching frequencies.

Table 6.3-6   Five-Segment Switching Sequence

Sector	Switching Sequence (Type A)					
I	$\vec{V}_0$	$\vec{V}_1$	$\vec{V}_2$	$\vec{V}_1$	$\vec{V}_0$	$v_{CN} = 0$
	OOO	POO	PPO	POO	OOO	
II	$\vec{V}_0$	$\vec{V}_3$	$\vec{V}_2$	$\vec{V}_3$	$\vec{V}_0$	$v_{CN} = 0$
	OOO	OPO	PPO	OPO	OOO	
III	$\vec{V}_0$	$\vec{V}_3$	$\vec{V}_4$	$\vec{V}_3$	$\vec{V}_0$	$v_{AN} = 0$
	OOO	OPO	OPP	OPO	OOO	
IV	$\vec{V}_0$	$\vec{V}_5$	$\vec{V}_4$	$\vec{V}_5$	$\vec{V}_0$	$v_{AN} = 0$
	OOO	OOP	OPP	OOP	OOO	
V	$\vec{V}_0$	$\vec{V}_5$	$\vec{V}_6$	$\vec{V}_5$	$\vec{V}_0$	$v_{BN} = 0$
	OOO	OOP	POP	OOP	OOO	
VI	$\vec{V}_0$	$\vec{V}_1$	$\vec{V}_6$	$\vec{V}_1$	$\vec{V}_0$	$v_{BN} = 0$
	OOO	POO	POP	POO	OOO	

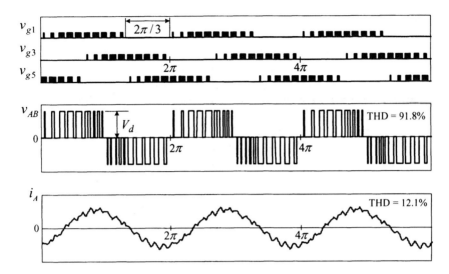

Figure 6.3-12 Waveforms produced by five-segment SVM with $f_1 = 60$ Hz, $f_{sw} = 600$ Hz, $T_s = 1/900$ s and $m_a = 0.8$.

6.4 SUMMARY

This chapter focuses on pulse-width modulation schemes for the two-level voltage source inverter. The switching frequency of the inverter is usually limited to a few hundred hertz for high-power medium-voltage (MV) drives. A carrier-based sinusoidal pulse-width modulation (SPWM) scheme is reviewed, followed by a detailed analysis on space vector modulation (SVM) algorithms, including derivation of space vectors, calculation of dwell times, design of switching sequence, and analysis of harmonic spectrum and THD.

The SVM schemes usually generate both odd- and even-order harmonics in the inverter output voltages. The even-order harmonics may not have a significantly impact on the operation of the motor. However, they are strictly regulated by harmonic guidelines such as IEEE Standard 519-1992 when the two-level converter is used as a rectifier in the MV drive. Since the two-level voltage source rectifier is not separately discussed in the book, the mechanism of even-order harmonic generation is analyzed and a modified SVM scheme for even-order harmonic eliminations presented.

The two-level inverter has a number of features, including simple converter topology and PWM scheme. However, the inverter produces high dv/dt and THD in its output voltage, and therefore often requires a large-size LC filter installed at its output terminals. Other advantages and drawbacks of the two-level inverter for use in the MV drive will be elaborated in Chapter 12.

REFERENCES

1. N. Mohan, T. M. Undeland, et al., *Power Electronics—Converters, Applications and Design,* 3rd edition, John Wiley & Sons, New York, 2003.
2. A. M. Hava, R. J. Kerkman, et al., Carrier-based PWM-VSI Overmodulation Strategies: Analysis, Comparison and Design, *IEEE Transactions on Power Electronics,* Vol. 13, No. 4, pp. 674–689, 1998.
3. D. G. Holmes and T. A. Lipo, *Pulse Width Modulation for Power Converters—Principle and Practice,* IEEE Press/Wiley-Interscience, New York, 2003.
4. M. H. Rashid, *Power Electronics Handbook,* Academic Press, New York, 2001.
5. P. C. Krause, O. Wasynczuk, et al., *Analysis of Electric Machinery and Drive Systems,* 2nd edition, IEEE Press/Wiley-Interscience, New York, 2002.

Chapter 7

Cascaded H-Bridge Multilevel Inverters

7.1 INTRODUCTION

Cascaded H-bridge (CHB) multilevel inverter is one of the popular converter topologies used in high-power medium-voltage (MV) drives [1–3]. It is composed of a multiple units of single-phase H-bridge power cells. The H-bridge cells are normally connected in cascade on their ac side to achieve medium-voltage operation and low harmonic distortion. In practice, the number of power cells in a CHB inverter is mainly determined by its operating voltage and manufacturing cost. For instance, in the MV drives with a rated line-to-line voltage of 3300 V, a nine-level inverter can be used, where the CHB inverter has a total of 12 power cells using 600 V class components [1]. The use of identical power cells leads to a modular structure, which is an effective means for cost reduction.

The CHB multilevel inverter requires a number of isolated dc supplies, each of which feeds an H-bridge power cell. The dc supplies are normally obtained from multipulse diode rectifiers presented in Chapter 3. For the seven- and nine-level inverters, 18- and 24-pulse diode rectifiers can be employed, respectively, to achieve low line-current harmonic distortion and high input power factor.

In this chapter, the single-phase H-bridge power cell, which is the building block for the CHB inverter, is reviewed. Various inverter topologies are introduced. Two carrier-based PWM schemes, phase-shifted and level-shifted modulations, are analyzed and their performance is compared. A staircase modulation with selective harmonic elimination is also presented.

7.2 H-BRIDGE INVERTER

Figure 7.2-1 shows a simplified circuit diagram of a single-phase H-bridge inverter. It is composed of two **inverter legs** with two IGBT devices in each leg. The invert-

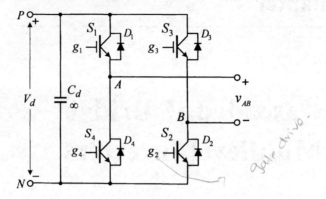

Figure 7.2-1 Single-phase H-bridge inverter.

er dc bus voltage V_d is usually fixed, while its ac output voltage v_{AB} can be adjusted by either bipolar or unipolar modulation schemes.

7.2.1 Bipolar Pulse-Width Modulation

Figure 7.2-2 shows a set of typical waveforms of the H-bridge inverter with bipolar modulation, where v_m is the sinusoidal modulating wave, v_{cr} is the triangular carrier wave, and v_{g1} and v_{g3} are the gate signals for the upper switches S_1 and S_3, respectively. The upper and the lower switches in the same inverter leg operate in a complementary manner with one switch turned on and the other turned off. Thus, we only need to consider two independent gate signals, v_{g1} and v_{g3}, which are generated by comparing v_m with v_{cr}. Following the same procedures given in Chapter 6, the waveforms of the inverter terminal voltages v_{AN} and v_{BN} can be derived, from which the inverter output voltage can be found from $v_{AB} = v_{AN} - v_{BN}$. Since the waveform of v_{AB} switches between the positive and negative dc voltages $\pm V_d$, this scheme is known as **bipolar modulation** [4].

The harmonic spectrum of the inverter output voltage v_{AB} normalized to its dc voltage V_d is shown in Fig. 7.2-2b, where V_{ABn} is the rms value of the nth-order harmonic voltage. The harmonics appear as sidebands centered around the frequency modulation index m_f and its multiples such as $2m_f$ and $3m_f$. The voltage harmonics with the order lower than $(m_f - 2)$ are either eliminated or negligibly small. The switching frequency of the IGBT device, referred to as **device switching frequency** $f_{sw,dev}$, is equal to the carrier frequency f_{cr}.

Figure 7.2-3 shows the harmonic content of v_{AB} versus the amplitude modulation index m_a. The fundamental voltage V_{AB1} (rms) increases linearly with m_a. The dominant harmonic m_f has a high magnitude, which is even higher than V_{AB1} for $m_a <$ 0.8. This harmonic along with its sidebands can be eliminated by the unipolar pulse width modulation scheme.

7.2 H-Bridge Inverter

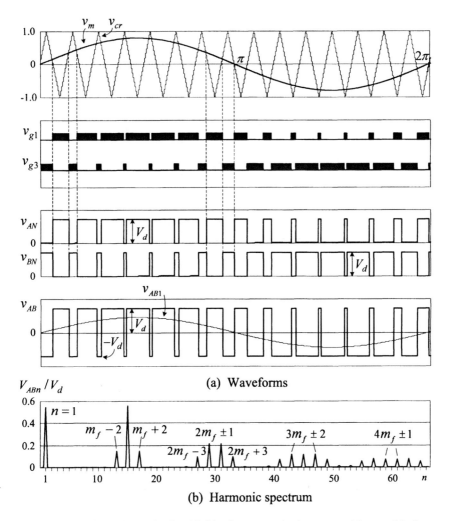

Figure 7.2-2 Bipolar PWM for the H-bridge inverter operating at $m_f = 15$, $m_a = 0.8$, $f_m = 60$ Hz, and $f_{cr} = 900$ Hz.

7.2.2 Unipolar Pulse-Width Modulation

The unipolar modulation normally requires two sinusoidal modulating waves, v_m and v_{m-}, which are of the same magnitude and frequency but 180° out of phase as shown in Fig. 7.2-4. The two modulating waves are compared with a common triangular carrier wave v_{cr}, generating two gating signals, v_{g1} and v_{g3}, for the upper switches, S_1 and S_3, respectively. It can be observed that the two upper devices do not switch simultaneously, which is distinguished from the bipolar PWM where all four devices are switched at the same time. The inverter output voltage v_{AB} switches

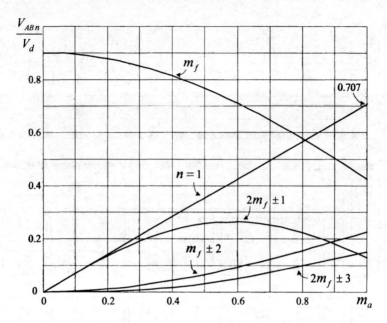

Figure 7.2-3 Harmonic content of v_{AB} produced by the H-bridge inverter with bipolar PWM.

either between zero and $+V_d$ during the positive half-cycle or between zero and $-V_d$ during the negative half-cycle of the fundamental frequency. Thus, this scheme is known as **unipolar modulation** [4].

Figure 7.2-4b shows the harmonic spectrum of the inverter output voltage v_{AB}. The harmonics appear as sidebands centered around $2m_f$ and $4m_f$. The low-order harmonics generated by the bipolar modulation, such as m_f and $m_f \pm 2$, are eliminated by the unipolar modulation. The dominant harmonics are distributed around $2m_f$, and their frequencies are in the neighborhood of 1800 Hz. This is essentially the equivalent **inverter switching frequency** $f_{sw,inv}$, which is also the switching frequency seen by the load. Compared with the device switching frequency of 900 Hz, the inverter switching frequency is doubled. This phenomenon can also be explained from another perspective. The H-bridge inverter has two complementary switch pairs switching at 900 Hz. But the two pairs normally switch at different time instants, leading to $f_{sw,inv} = 2f_{sw,dev}$.

It is interesting to note that the dominant harmonics, $2m_f \pm 1$ and $2m_f \pm 3$, produced by the unipolar modulation have exactly the same magnitude as those generated by bipolar modulation. As a result, Fig. 7.2-3 can also be used to determine the magnitude of these harmonics at various m_a.

The unipolar modulation can also be implemented by using only one modulating wave v_m but two phase-shifted carrier waves, v_{cr} and v_{cr-}, as shown in Fig. 7.2-5. The two carrier waves are of same amplitude and frequency, but 180° out of phase. Switch S_1 is turned on by v_{g1} when $v_m > v_{cr}$, whereas S_3 is on when $v_m < v_{cr-}$. The

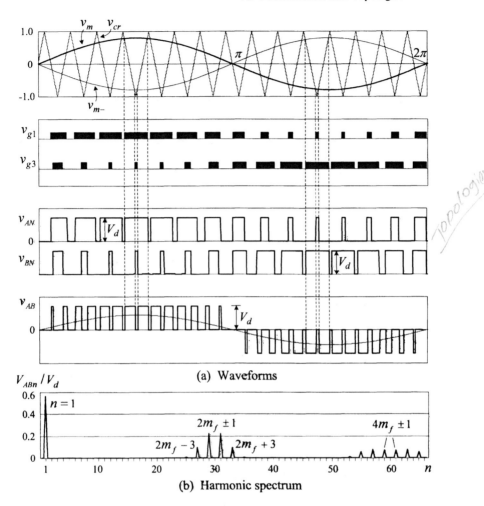

Figure 7.2-4 Unipolar PWM with two phase-shifted modulating waves ($m_f = 15$, $m_a = 0.8$, $f_m = 60$ Hz, and $f_{cr} = 900$ Hz).

waveform of v_{AB} is identical to that shown in Fig. 7.2-4. This modulation technique is often used in the CHB multilevel inverters.

7.3 MULTILEVEL INVERTER TOPOLOGIES

7.3.1 CHB Inverter with Equal dc Voltage

As the name suggests, the cascaded H-bridge multilevel inverter uses multiple units of H-bridge power cells connected in a series chain to produce high ac voltages. A typical configuration of a five-level CHB inverter is shown in Fig. 7.3-1, where

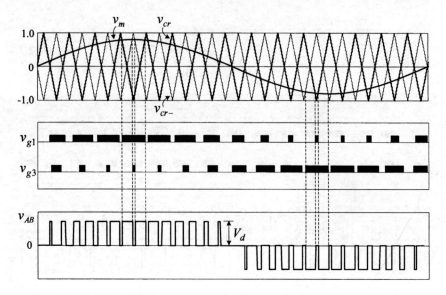

Figure 7.2-5 Unipolar PWM with two phase-shifted carriers ($m_f = 15$, $m_a = 0.8$, $f_m = 60$, and $f_{cr} = 900$ Hz).

each phase leg consists of two H-bridge cells powered by two isolated dc supplies of equal voltage E. The dc supplies are normally obtained by multipulse diode rectifiers discussed in Chapter 3.

The CHB inverter in Fig. 7.3-1 can produce a phase voltage with five voltage levels. When switches S_{11}, S_{21}, S_{12}, and S_{22} conduct, the output voltage of the H-bridge cells $H1$ and $H2$ is $v_{H1} = v_{H2} = E$, and the resultant **inverter phase voltage** is $v_{AN} = v_{H1} + v_{H2} = 2E$, which is the voltage at the inverter terminal A with respect to the inverter neutral N. Similarly, with S_{31}, S_{41}, S_{32}, and S_{42} switched on, $v_{AN} = -2E$. The other three voltage levels are E, 0, and $-E$, which correspond to various switching states summarized in Table 7.3-1. It is worth noting that the inverter phase voltage v_{AN} may not necessarily equal the **load phase voltage** v_{AO}, which is the voltage at node A with respect to the load neutral O.

It can be observed from Table 7.3-1 that some voltage levels can be obtained by more that one switching state. The voltage level E, for instance, can be produced by four sets of different (redundant) switching states. The switching state redundancy is a common phenomenon in multilevel converters. It provides a great flexibility for switching pattern design, especially for space vector modulation schemes.

The number of voltage levels in a CHB inverter can be found from

$$m = (2H + 1) \qquad (7.3\text{-}1)$$

where H is the number of H-bridge cells per phase leg. The voltage level m is always an odd number for the CHB inverter while in other multilevel topologies such as diode-clamped inverters, it can be either an even or odd number.

7.3 Multilevel Inverter Topologies

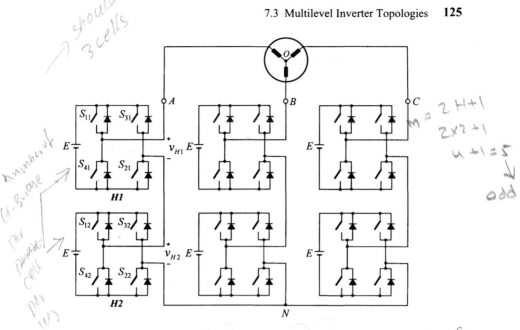

Figure 7.3-1 Five-level cascaded H-bridge inverter.

Table 7.3-1 Voltage Level and Switching State of the Five-Level CHB Inverter

Output Voltage v_{AN}	Switching State				v_{H1}	v_{H2}
	S_{11}	S_{31}	S_{12}	S_{32}		
$2E$	1	0	1	0	E	E
E	1	0	1	1	E	0
	1	0	0	0	E	0
	1	1	1	0	0	E
	0	0	1	0	0	E
0	0	0	0	0	0	0
	0	0	1	1	0	0
	1	1	0	0	0	0
	1	1	1	1	0	0
	1	0	0	1	E	$-E$
	0	1	1	0	$-E$	E
$-E$	0	1	1	1	$-E$	0
	0	1	0	0	$-E$	0
	1	1	0	1	0	$-E$
	0	0	0	1	0	$-E$
$-2E$	0	1	0	1	$-E$	$-E$

The CHB inverter introduced above can be extended to any number of voltage levels. The per-phase diagram of seven- and nine-level inverters are depicted in Fig. 7.3-2, where the seven-level inverter has three H-bridge cells in cascade while the nine-level has four cells in series. The total number of **active switches** (IGBTs) used in the CHB inverters can be calculated by

$$N_{sw} = 6(m-1) \tag{7.3-2}$$

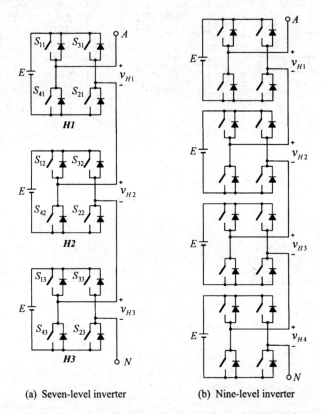

(a) Seven-level inverter   (b) Nine-level inverter

Figure 7.3-2 Per-phase diagram of seven- and nine-level CHB inverters.

7.3.2 H-Bridges with Unequal dc Voltages

The dc supply voltages of the H-bridge power cells introduced in the previous section are all the same. Alternatively, different dc voltages may be selected for the power cells. With unequal dc voltages, the number of voltage levels can be increased without necessarily increasing the number of H-bridge cells in cascade. This allows more voltage steps in the inverter output voltage waveform for a given number of power cells [5, 6].

Figure 7.3-3 shows two inverter topologies, where the dc voltages for the H-bridge cells are not equal. In the seven-level topology, the dc voltages for *H1* and

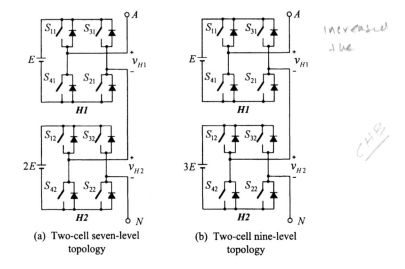

Figure 7.3-3 Per-phase diagram of CHB inverters with unequal dc voltages.

H2 are E and $2E$, respectively. The two-cell inverter leg is able to produce seven voltage levels: $3E$, $2E$, E, 0, $-E$, $-2E$, and $-3E$. The relationship between the voltage levels and their corresponding switching states is summarized in Table 7.3-2. In the nine-level topology, the dc voltage of H2 is three times that of H1. All the nine voltage levels can be obtained by replacing the H2 output voltage of $v_{H2} = \pm 2E$ in Table 7.3-2 with $v_{H2} = \pm 3E$ and then calculating the inverter phase voltage v_{AN}.

There are some drawbacks associated with the CHB inverter using unequal dc voltages. The merits of the modular structure are essentially lost. In addition, switching pattern design becomes much more difficult due to the reduction in redundant switching states [5]. Therefore, this inverter topology has limited industrial applications.

7.4 CARRIER-BASED PWM SCHEMES

The carrier-based modulation schemes for multilevel inverters can be generally classified into two categories: **phase-shifted** and **level-shifted modulations**. Both modulation schemes can be applied to the CHB inverters.

7.4.1 Phase-Shifted Multicarrier Modulation

In general, a multilevel inverter with m voltage levels requires $(m-1)$ triangular carriers. In the phase-shifted multicarrier modulation, all the triangular carriers have the same frequency and the same peak-to-peak amplitude, but there is a phase shift between any two adjacent carrier waves, given by

$$\phi_{cr} = 360°/(m-1) \qquad (7.4\text{-}1)$$

Table 7.3-2 Voltage Level and Switching State of the Two-Cell Seven-Level CHB Inverter with Unequal dc Voltages

Output Voltage v_{AN}	Switching State				v_{H1}	v_{H2}
	S_{11}	S_{31}	S_{12}	S_{32}		
$3E$	1	0	1	0	E	$2E$
$2E$	1	1	1	0	0	$2E$
	0	0	1	0	0	$2E$
E	1	0	1	1	E	0
	1	0	0	0	E	0
	0	1	1	0	$-E$	$2E$
0	0	0	0	0	0	0
	0	0	1	1	0	0
	1	1	0	0	0	0
	1	1	1	1	0	0
$-E$	1	0	0	1	E	$-2E$
	0	1	1	1	$-E$	0
	0	1	0	0	$-E$	0
$-2E$	1	1	0	1	0	$-2E$
	0	0	0	1	0	$-2E$
$-3E$	0	1	0	1	$-E$	$-2E$

The modulating signal is usually a three-phase sinusoidal wave with adjustable amplitude and frequency. The gate signals are generated by comparing the modulating wave with the carrier waves.

Figure 7.4-1 shows the principle of the phase-shifted modulation for a seven-level CHB inverter, where six triangular carriers are required with a 60° phase displacement between any two adjacent carriers. Of the three-phase sinusoidal modulating waves, only the phase A modulating wave v_{mA} is plotted for simplicity. The carriers v_{cr1}, v_{cr2}, and v_{cr3} are used to generate gatings for the upper switches S_{11}, S_{12}, and S_{13} in the left legs of power cells $H1$, $H2$, and $H3$ in Fig. 7.3-2a, respectively. The other three carriers, v_{cr1-}, v_{cr2-} and v_{cr3-}, which are 180° out of phase with v_{cr1}, v_{cr2} and v_{cr3}, respectively, produce the gatings for the upper switches S_{31}, S_{32}, and S_{33} in the right legs of the H-bridge cells. The gate signals for all the lower switches in the H-bridge legs are not shown since these switches operate in a complementary manner with respect to their corresponding upper switches.

The PWM scheme discussed above is essentially the unipolar modulation. As shown in Fig. 7.4-1, the gatings for the upper switches S_{11} and S_{31} in $H1$ are generated by comparing v_{cr1} and v_{cr1-} with v_{mA}. The $H1$ output voltage v_{H1} is switched either between zero and E during the positive half-cycle or between zero and $-E$ during the negative half-cycle of the fundamental frequency. The frequency modulation index in this example is $m_f = f_{cr}/f_m = 3$ and the amplitude modulation index is $m_a = \hat{V}_{mA}/\hat{V}_{cr} = 0.8$, where f_{cr} and f_m are the frequencies of the carrier and modulating waves, and $\hat{V}_{mA}$ and $\hat{V}_{cr}$ are the peak amplitudes of v_{mA} and v_{cr}, respectively.

7.4 Carrier-Based PWM Schemes

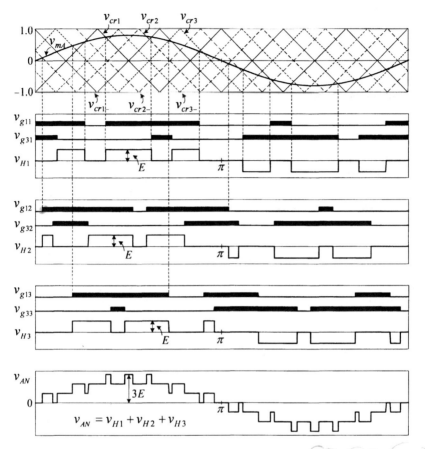

Figure 7.4-1 Phase-shifted PWM for seven-level CHB inverters ($m_f = 3$, $m_a = 0.8$, $f_m = 60$ Hz, and $f_{cr} = 180$ Hz).

The inverter phase voltage can be found from

$$v_{AN} = v_{H1} + v_{H2} + v_{H3} \tag{7.4-2}$$

where v_{H1}, v_{H2}, and v_{H3} are the output voltages of cells H1, H2, and H3, respectively. It is clear that the inverter phase voltage waveform is formed by seven voltage steps: $+3E$, $2E$, E, 0, $-E$, $-2E$, and $-3E$.

Figure 7.4-2 shows the simulated voltage waveforms and their harmonic content of the seven-level inverter operating under the condition of $f_m = 60$ Hz, $m_f = 10$, and $m_a = 1.0$. The device switching frequency can be calculated by $f_{sw,dev} = f_{cr} = f_m \times m_f = 600$ Hz, which is a typical value for the switching devices in high-power converters. The waveforms of v_{H1}, v_{H2}, and v_{H3} are almost identical except for a small phase displacement caused by the phase-shifted carriers.

The waveform of v_{AN} is composed of seven voltage levels with a peak value of $3E$. Since the IGBTs in the different H-bridges do not switch simultaneously, the magnitude of voltage step change during switching is only E. This leads to a low dv/dt and reduced electromagnetic interference (EMI). The line-to-line voltage v_{AB} has 13 voltage levels with an amplitude of $6E$.

The harmonic spectrum for the waveforms of v_{H1}, v_{AN}, and v_{AB} is shown in Fig. 7.4-2b. The harmonics in v_{H1} appear as sidebands centered around $2m_f$ and its multiples such as $4m_f$ and $6m_f$. The harmonic content of v_{H2} and v_{H3} is identical to that of v_{H1}, and thus it is not given in the figure. The inverter phase voltage v_{AN} does not contain any harmonics of the order lower than $4m_f$, which leads to a significant reduction in THD. The THD for v_{AN} is only 18.8% in comparison to 53.9% for v_{H1}. It can be observed that v_{AN} contains triplen harmonics such as $(6m_f \pm 3)$ and $(6m_f \pm 9)$. However, these harmonics do not appear in the line-to-line voltage v_{AB} due to the balanced three-phase system, resulting in a further reduction in THD to 15.6%.

As stated earlier, the frequency of the dominant harmonic in the inverter output voltage represents the inverter switching frequency $f_{sw,inv}$. Since the dominant harmonics in v_{AN} and v_{AB} in Fig. 7.4-2 are distributed around $6m_f$, the inverter switching frequency can be found from $f_{sw,inv} = 6m_f \times f_m = 6f_{sw,dev}$, which is six times the device switching frequency. This is a desirable feature attained by the multilevel inverter since a high value of $f_{sw,inv}$ allows more harmonics in v_{AB} to be eliminated while a low value of $f_{sw,dev}$ helps to reduce device switching losses. In general, the switching frequency of the inverter using the phase-shifted modulation is related to the device switching frequency by

$$f_{sw,inv} = 2Hf_{sw,dev} = (m-1)f_{sw,dev} \qquad (7.4\text{-}3)$$

The harmonic content of v_{AB} versus the modulation index m_a is shown in Fig. 7.4-3. Since the high-order harmonic components can be easily attenuated by filters or load inductances, only the dominant harmonics centered around $6m_f$ are plotted. The nth-order harmonic voltage V_{ABn} (rms) is normalized with respect to the total dc voltage

$$V_d = \frac{m-1}{2} E \qquad (7.4\text{-}4)$$

For a seven-level inverter, $V_d = 3E$. The maximum fundamental-frequency voltage can be found from

$$V_{AB1,max} = 1.224 V_d = 0.612(m-1)E \qquad \text{for } m_a = 1 \qquad (7.4\text{-}5)$$

As discussed in Chapter 6, the maximum voltage $V_{AB1,max}$ can be boosted by 15.5% by the **third harmonic injection method**. This technique can also be applied to the phase- and level-shifted modulation schemes for the CHB inverters.

7.4 Carrier-Based PWM Schemes

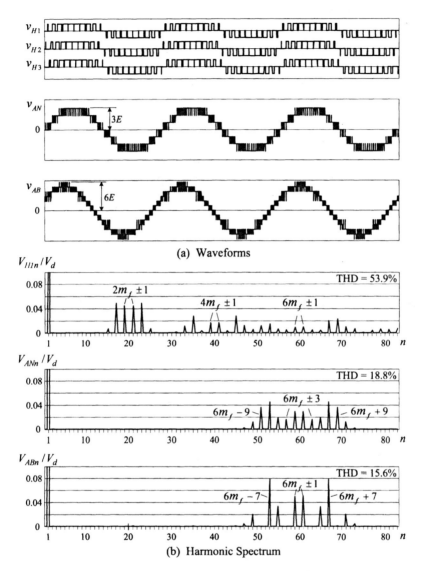

Figure 7.4-2 Simulated waveforms for a seven-level CHB inverter with phase-shifted PWM ($m_f = 10$, $m_a = 1.0$, $f_m = 60$ Hz, and $f_{cr} = 600$ Hz).

7.4.2 Level-Shifted Multicarrier Modulation

Similar to the phase-shifted modulation, an m-level CHB inverter using level-shifted multicarrier modulation scheme requires $(m-1)$ triangular carriers, all having the same frequency and amplitude. The $(m-1)$ triangular carriers are vertically disposed such that the bands they occupy are contiguous. The frequency modulation

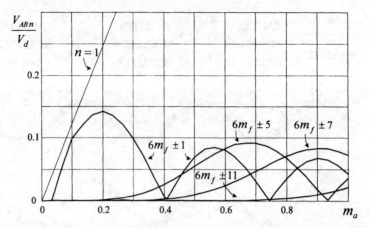

Figure 7.4-3 Harmonic content of v_{AB} produced by a seven-level CHB inverter with phase-shifted PWM.

index is given by $m_f = f_{cr}/f_m$, which remains the same as that for the phase-shifted modulation scheme whereas the amplitude modulation index is defined as

$$m_a = \frac{\hat{V}_m}{\hat{V}_{cr}(m-1)} \quad \text{for } 0 \le m_a \le 1 \quad (7.4\text{-}6)$$

where $\hat{V}_m$ is the peak amplitude of the modulating wave v_m and $\hat{V}_{cr}$ is the peak amplitude of each carrier wave.

Figure 7.4-4 shows three schemes for the level-shifted multicarrier modulation: (a) in-phase disposition (IPD), where all carriers are in phase; (b) alternative phase opposite disposition (APOD), where all carriers are alternatively in opposite disposition; and (c) phase opposite disposition (POD), where all carriers above the zero reference are in phase but in opposition with those below the zero reference. In what follows, only IPD modulation scheme is discussed since it provides the best harmonic profile of all three modulation schemes [7].

Figure 7.4-5 shows the principle of the IPD modulation for a seven-level CHB inverter operating under the condition of $m_f = 15$, $m_a = 0.8$, $f_m = 60$ Hz, and $f_{cr} = f_m \times m_f = 900$ Hz. The uppermost and lowermost carrier pair, v_{cr1} and v_{cr1-}, are used to generate the gatings for switches S_{11} and S_{31} in power cell $H1$ of Fig. 7.3-2a. The innermost carrier pair, v_{cr3} and v_{cr3-}, generate gatings for S_{13} and S_{33} in $H3$. The remaining carrier pair, v_{cr2} and v_{cr2-}, are for S_{12} and S_{32} in $H2$. For the carriers above the zero reference (v_{cr1}, v_{cr2}, and v_{cr3}), the switches S_{11}, S_{12}, and S_{13} are turned on when the phase A modulating signal v_{mA} is higher than the corresponding carriers. For the carriers below the zero reference (v_{cr1-}, v_{cr2-}, and v_{cr3-}), S_{31}, S_{32}, and S_{33} are switched on when v_{mA} is lower than the carrier waves. The gate signals for the lower switches in each H-bridge are complementary to their corresponding upper switches, and thus for simplicity they are not shown. The resultant H-bridge output voltage waveforms v_{H1}, v_{H2}, and v_{H3} are all unipolar as shown in Fig. 7.4-5. The inverter phase voltage waveform v_{AN} is formed with seven voltage levels.

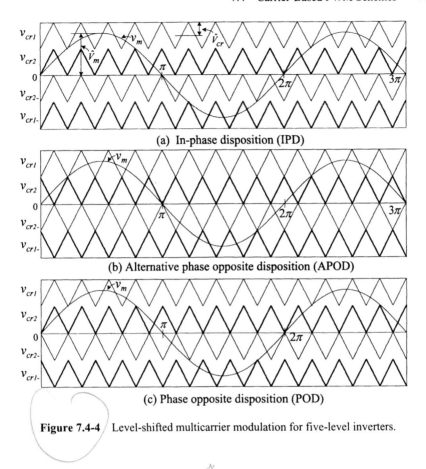

(a) In-phase disposition (IPD)

(b) Alternative phase opposite disposition (APOD)

(c) Phase opposite disposition (POD)

Figure 7.4-4 Level-shifted multicarrier modulation for five-level inverters.

In the phase-shifted modulation, the device switching frequency is equal to the carrier frequency. This relationship, however, is no longer held true for the IPD modulation. For example, with the carrier frequency of 900 Hz in Fig. 7.4-5, the switching frequency of the devices in $H1$ is only 180 Hz, which is obtained by the number of gating pulses per cycle multiplied by the frequency of the modulating wave (60 Hz). Furthermore, the switching frequency is not the same for the devices in different H-bridge cells. The switches in $H3$ are turned on and off only once per cycle, which translates into a switching frequency of 60 Hz. In general, the switching frequency of the inverter using the level-shifted modulation is equal to the carrier frequency, that is,

$$f_{sw,inv} = f_{cr} \qquad (7.4\text{-}7)$$

from which the average device switching frequency is

$$f_{sw,dev} = f_{cr}/(m-1) \qquad (7.4\text{-}8)$$

In addition to the unequal device switching frequencies, the conduction time of the

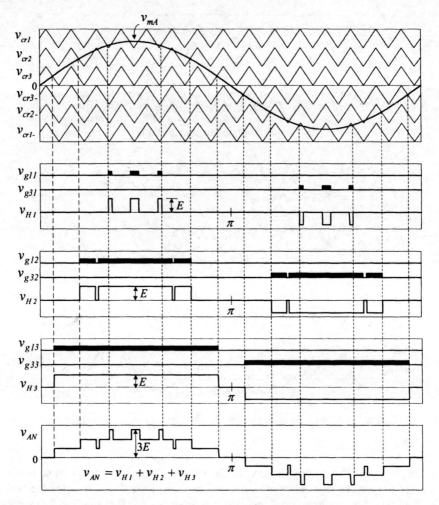

Figure 7.4-5 Level-shifted PWM for a seven-level CHB inverter ($m_f = 15$, $m_a = 0.8$, $f_m = 60$ Hz, and $f_{cr} = 900$ Hz).

devices is not evenly distributed either. For example, the device S_{11} in *H1* conduct much less time than S_{13} in *H3* per cycle of the fundamental frequency. To evenly distribute the switching and conduction losses, the switching pattern should rotate among the H-bridge cells.

Figure 7.4-6 shows the simulated waveforms for a seven-level inverter operating under the condition of $m_f = 60$, $m_a = 1.0$, $f_m = 60$ Hz, and $f_{cr} = 3600$ Hz. Although the carrier frequency of 3600 Hz seems high for high-power converters, the average device switching frequency is only 600 Hz. The output voltages of the H-bridge cells, v_{H1}, v_{H2}, and v_{H3}, are all different, signifying that the IGBTs operate at different switching frequencies with various conduction times.

Similar to the voltage waveforms produced by the phase-shifted modulation,

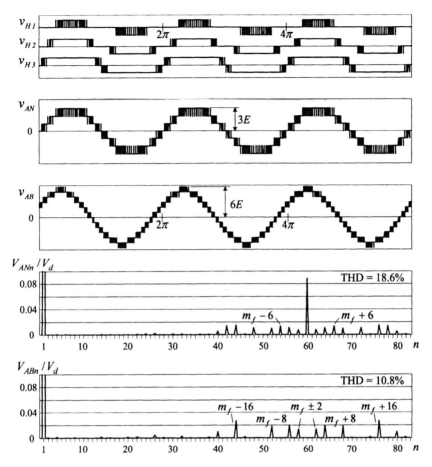

Figure 7.4-6 Simulated waveforms for a seven-level CHB inverter with IPD modulation ($m_f = 60$, $m_a = 1.0$, $f_m = 60$ Hz, $f_{cr} = 3600$ Hz, and $f_{sw,dev} = 600$ Hz).

the inverter phase voltage v_{AN} is composed of seven voltage levels while the line-to-line voltage v_{AB} has 13 voltage levels. The dominant harmonics in v_{AN} and v_{AB} appear as sidebands centered around m_f. The inverter phase voltage contains triplen harmonics, such as m_f and $m_f \pm 6$, with m_f being a dominant harmonic. Since these harmonics do not appear in the line-to-line voltage, the THD of v_{AB} is only 10.8% in comparison to 18.6% for v_{AN}. The spectra of v_{AB} at other modulation indices m_a are shown in Fig. 7.4-7. The THD of v_{AB} decreases from 48.8% at $m_a = 0.2$ to 13.1% at $m_a = 0.8$.

The waveforms for v_{AN} and v_{AB} measured from a laboratory seven-level CHB inverter are illustrated in Fig. 7.4-8. The inverter operates under the condition of $m_f = 60$, $m_a = 1.0$, $f_m = 60$ Hz, and $f_{cr} = 3600$ Hz. The measured waveforms and their harmonic spectra are consistent with the simulation results shown in Fig. 7.4-6.

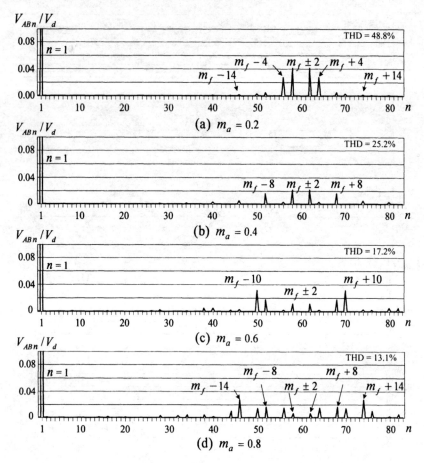

Figure 7.4-7 Harmonic content of v_{AB} produced by a seven-level CHB inverter with IPD modulation ($m_f = 60$, $f_m = 60$ Hz, $f_{cr} = 3600$ Hz, and $f_{sw,dev} = 600$ Hz).

7.4.3 Comparison Between Phase- and Level-Shifted PWM Schemes

To compare the performance of phase- and level-shifted modulation schemes, it is assumed that the average switching frequency of the solid-state devices is the same for both schemes. Figure 7.4-9 shows the output voltage waveforms of a seven-level inverter operating with $f_{sw,dev} = 600$ Hz and $m_a = 0.2$, at which the differences between the two modulation schemes can be easily distinguished.

The H-bridge output voltages, v_{H1}, v_{H2}, and v_{H3}, produced by the phase-shifted modulation are almost identical except a small phase displacement among them. All the devices operate at the same switching frequency and conduction time. However, v_{H1} and v_{H2} produced by the level-shifted modulation are equal to zero and thus no

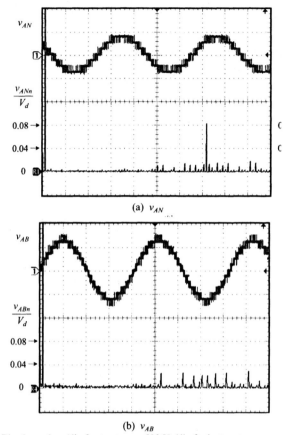

(b) v_{AB}
Timebase: 4 ms/div for top traces; 500 Hz/div for bottom traces.

Figure 7.4-8 Waveforms measured from a laboratory seven-level CHB inverter with IPD modulation ($m_f = 60$, $m_a = 1.0$, $f_m = 60$ Hz, $f_{cr} = 3600$ Hz, and $f_{sw,dev} = 600$ Hz).

switchings occur in power cells $H1$ and $H2$. The devices in $H3$ switch at the carrier frequency of 3600 Hz. To evenly distribute the switching and conduction losses, the switching pattern for the devices in the H-bridge cells should rotate.

The inverter phase voltage v_{AN} produced by both modulation schemes looks similar. It contains only three voltage levels instead of seven due to the low modulation index. The voltage levels of the inverter line-to-line voltage v_{AB} are reduced accordingly. Furthermore, the THD of v_{AB} produced by the phase-shifted modulation is 96.7%, much higher than 48.8% for the level-shifted modulation. This is mainly caused by the waveform differences in the center portion of the positive and negative half-cycles of v_{AB}.

Figure 7.4-10 shows the THD profile of the line-to-line voltage v_{AB} modulated by the phase- and level-shifted schemes. A summary of the carrier-based modulation schemes for the CHB multilevel inverters is given in Table 7.4-1.

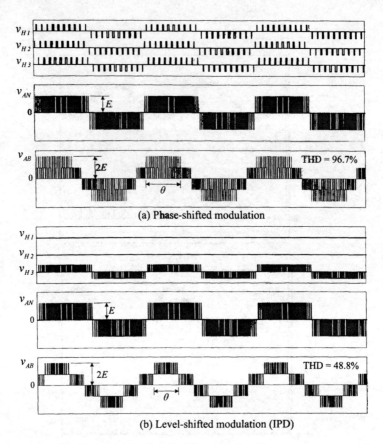

Figure 7.4-9 Output voltage waveforms of the seven-level inverter operating at a low modulation index.

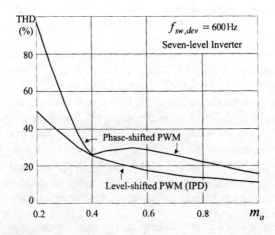

Figure 7.4-10 THD profile of v_{AB} produced by the seven-level CHB inverter with phase- and level-shifted modulation schemes.

Table 7.4-1 Comparison Between the Phase- and Level-Shifted PWM Schemes

Comparison	Phase-Shifted Modulation	Level-Shifted Modulation (IPD)
Device switching frequency	Same for all devices	Different
Device conduction period	Same for all devices	Different
Rotating of switching patterns	No required	Required
Line-to-line voltage THD	Good	Better

7.5 STAIRCASE MODULATION

The staircase modulation can be easily implemented for the CHB inverter due to its unique structure [8, 9]. The principle of this modulation scheme is illustrated in Fig. 7.5-1, where v_{H1}, v_{H2}, and v_{H3} are the output voltages of the H-bridge cells in a seven-level inverter shown in Fig. 7.3-2a. The inverter phase voltage v_{AN} is formed by a seven-level staircase.

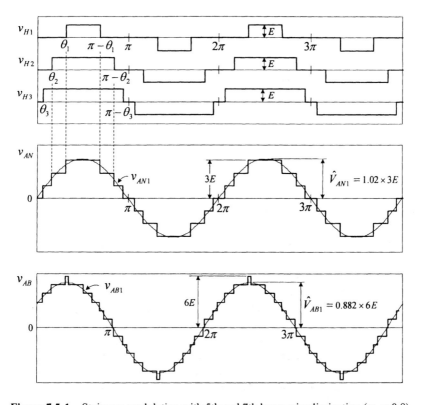

Figure 7.5-1 Staircase modulation with 5th and 7th harmonic elimination ($m_a = 0.8$).

The waveform of v_{AN} can be expressed in terms of Fourier series as

$$v_{AN} = \frac{4E}{\pi} \sum_{n=1,3,5,\ldots}^{\infty} \frac{1}{n} \{\cos(n\theta_1) + \cos(n\theta_2) + \cos(n\theta_3)\} \sin(n\omega t) \quad (7.5\text{-}1)$$

$$\text{for } 0 \leq \theta_3 < \theta_2 < \theta_1 \leq \pi/2$$

where n is the harmonic order, and θ_1, θ_2 and θ_3 are the independent switching angles. The coefficient $4E/\pi$ represents the peak value of the maximum fundamental voltage $\hat{V}_{H,\max}$ of an H-bridge cell, which occurs when the switching angle θ_1 of v_{H1}, for example, reduces to zero.

The three independent angles can be used to eliminate two harmonics in v_{AN} and also provide an adjustable modulation index, defined by

$$m_a = \frac{\hat{V}_{AN1}}{H \times \hat{V}_{H,\max}} = \frac{\hat{V}_{AN1}}{H \times 4E/\pi} \quad (7.5\text{-}2)$$

where $\hat{V}_{AN1}$ is the peak value of the fundamental inverter phase voltage v_{AN1} and H is the number of H-bridge cells per phase.

For the seven-level CHB inverter with 5th and 7th harmonic elimination, the following equations can be formulated:

$$\cos(\theta_1) + \cos(\theta_2) + \cos(\theta_3) = 3m_a$$
$$\cos(5\theta_1) + \cos(5\theta_2) + \cos(5\theta_3) = 0 \quad (7.5\text{-}3)$$
$$\cos(7\theta_1) + \cos(7\theta_2) + \cos(7\theta_3) = 0$$

from which

$$\theta_1 = 57.106°, \ \theta_2 = 28.717° \text{ and } \theta_3 = 11.504° \quad \text{for } m_a = 0.8 \quad (7.5\text{-}4)$$

The inverter output voltage waveforms based on (7.5-4) are shown in Fig. 7.5-1, and their spectrum is illustrated in Fig. 7.5-2. The waveform of v_{AN} does not contain the 5th or 7th harmonics, and its THD is 12.5%. The inverter line-to-line voltage v_{AB} does not have any triplen harmonics such as 3rd, 9th, and 15th, resulting in a further reduction in THD.

The staircase modulation scheme is simple to implement. All the switching angles can be calculated off-line and then stored in a look-up table for digital implementation. Compared with the carrier-based PWM schemes, the staircase modulation features low switching losses since all the IGBTs operate at the fundamental frequency.

It is worth noting that the equations such as (7.5-3) for the switching angle calculation are nonlinear and transcendental, and thus they may not always have a valid solution over the full range of m_a [10]. When it happens, the switching angles should be calculated to minimize the magnitude of those harmonics that cannot be eliminated.

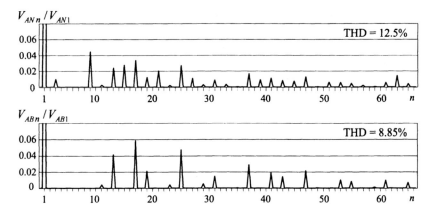

Figure 7.5-2 Harmonic spectrum for the waveforms of v_{AN} and v_{AB} in Fig. 7.5-1.

7.6 SUMMARY

This chapter focuses on the configurations and modulation schemes for cascaded H-bridge (CHB) multilevel inverters. The inverter is mainly composed of a number of identical H-bridge power cells connected in cascade. In practice, the number of H-bridge cells in a CHB inverter is primarily determined by the inverter operating voltage, harmonic requirements, and manufacturing cost. The CHB inverter with seven to eleven voltage levels has been increasingly used in high-power medium-voltage (MV) drives, where the IGBTs are exclusively used as switching devices.

Two multicarrier based PWM schemes, the phase- and level-shifted modulations, are presented. Various aspects associated with the modulation schemes for the CHB multilevel inverters are discussed, which include gate signal arrangements, spectrum analyses, and THD profiles. The performance of the modulation schemes is compared. Another commonly used modulation technique, space vector modulation, is not discussed in this chapter. The reader can refer to Chapters 6 and 8 for detailed analysis of the space vector modulation schemes.

The CHB multilevel inverter has a number of features and drawbacks, including

- **Modular structure.** The multilevel inverter is composed of multiple units of identical H-bridge power cells, which leads to a reduction in manufacturing cost;
- **Lower voltage THD and *dv/dt*.** The inverter output voltage waveform is formed by several voltage levels with small voltage steps. Compared with a two-level inverter, the CHB multilevel inverter can produce an output voltage with much lower THD and *dv/dt*;
- **High-voltage operation without switching devices in series.** The H-bridge

power cells are connected in cascade to produce high ac voltages. The problems of equal voltage sharing for series-connected devices are eliminated;

- **Large number of isolated dc supplies.** The dc supplies for the CHB inverter are usually obtained from a multipulse diode rectifier employing an expensive phase shifting transformer; and
- **High component count.** The CHB inverter uses a large number of IGBT modules. A nine-level CHB inverter requires 64 IGBTs with the same number of gate drivers.

REFERENCES

1. P. W. Hammond, A New Approach to Enhance Power Quality for Medium Voltage AC Drives, *IEEE Transactions on Industry Applications*, Vol. 33, No. 1, pp. 202–208, 1997.
2. W. A. Hill and C. D. Harbourt, Performance of Medium Voltage Multilevel Inverters, *IEEE Industry Applications Society (IAS) Conference*, Vol. 2, pp. 1186–1192, 1999.
3. R. H. Osman, A Medium Voltage Drive Utilizing Series-Cell Multilevel Topology for Outstanding Power Quality, *IEEE Industry Applications Society (IAS) Conference*, pp. 2662–2669, 1999.
4. N. Mohan, T. M. Undeland, et al., *Power Electronics—Converters, Applications and Design*, 3rd edition, John Wiley & Sons, New York, 2003.
5. P. W. Wheeler, L. Empringham, et al., Improved Output Waveform Quality for Multilevel H-Bridge Chain Converters Using Unequal Cell Voltages, *IEE Power Electronics and Variable Speed Drives Conference*, pp. 536–540, 2000.
6. M. D. Manjrekar, P.K. Steimer, et al., Hybrid Multilevel Power Conversion System: A Competitive Solution for High Power Applications, *IEEE Transactions on Industry Applications*, Vol. 36, No. 3, pp. 834–841, 2000.
7. G. Carrara, S. Gardella, et al., A New Multilevel PWM Method: A Theoretical Analysis, *IEEE Transactions on Power Electronics*, Vol. 7, No. 3, pp. 497–505, 1992.
8. L. M. Tolbert, F. Z. Peng, et al., Multilevel Converters for Large Electric Drives, *IEEE Transactions on Industry Applications*, Vol. 35, No. 1, pp. 36–44, 1999.
9. R. Kieferndorf, G. Venkataramanan, et al., A Power Electronic Transformer (PET) Fed Nine-Level H-Bridge Inverter for Large Induction Motor Drives, *IEEE Industry Applications Society (IAS) Conference*, Vol. 4, pp. 2489–2495, 2000.
10. J. Chiasson, L Tolbert, et al., Eliminating Harmonics in a Multilevel Converter Using Resultant Theory, *IEEE Power Electronics Specialists Conference (PESC)*, pp. 503–508, 2002.

Chapter 8

Diode Clamped Multilevel Inverters

8.1 INTRODUCTION

The diode-clamped multilevel inverter employs clamping diodes and cascaded dc capacitors to produce ac voltage waveforms with multiple levels. The inverter can be generally configured as a three-, four-, or five-level topology, but only the three-level inverter, often known as **neutral-point clamped** (NPC) inverter, has found wide application in high-power medium-voltage (MV) drives [1–3]. The main features of the NPC inverter include reduced dv/dt and THD in its ac output voltages in comparison to the two-level inverter discussed earlier. More importantly, the inverter can be used in the MV drive to reach a certain voltage level without switching devices in series. For instance, the NPC inverter using 6000-V devices is suitable for the drives rated at 4160 V.

In this chapter, various aspects of the three-level NPC inverter are discussed, including the inverter topology, operating principle and device commutation. A conventional **space vector modulation** (SVM) scheme for the NPC inverter is discussed in detail. To eliminate the even-order harmonics produced by the SVM, a modified modulation scheme is presented. The dc input voltage of the inverter is normally split by two cascaded dc capacitors, providing a floating neutral point. The control of the neutral-point voltage deviation is also elaborated. Finally, the operation of four- and five-level diode-clamped inverters with carrier-based modulation techniques is introduced.

8.2 THREE-LEVEL INVERTER

8.2.1 Converter Configuration

Figure 8.2-1 shows the simplified circuit diagram of a three-level NPC inverter. The inverter leg A is composed of four active switches S_1 to S_4 with four antiparallel diodes D_1 to D_4. In practice, either IGBT or GCT can be employed as a switching device.

144 Chapter 8 Diode-Clamped Multilevel Inverters

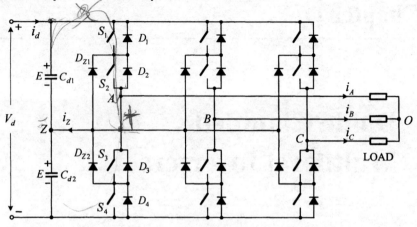

Figure 8.2-1 Three-level NPC inverter.

On the dc side of the inverter, the dc bus capacitor is split into two, providing a **neutral point** Z. The diodes connected to the neutral point, D_{Z1} and D_{Z2}, are the **clamping diodes**. When switches S_2 and S_3 are turned on, the inverter output terminal A is connected to the neutral point through one of the clamping diodes. The voltage across each of the dc capacitors is E, which is normally equal to half of the total dc voltage V_d. With a finite value for C_{d1} and C_{d2}, the capacitors can be charged or discharged by **neutral current** i_Z, causing **neutral-point voltage deviation**. This issue will be further discussed in the later sections.

8.2.2 Switching State

The operating status of the switches in the NPC inverter can be represented by switching states shown in Table 8.2-1. Switching state 'P' denotes that the upper two switches in leg A are on and the **inverter terminal voltage** v_{AZ}, which is the voltage at terminal A with respect to the neutral point Z, is $+E$, whereas 'N' indicates that the lower two switches conduct, leading to $v_{AZ} = -E$.

Switching state 'O' signifies that the inner two switches S_2 and S_3 are on and v_{AZ} is clamped to zero through the clamping diodes. Depending on the direction of load

Table 8.2-1 Definition of Switching States

Switching State	Device Switching Status (Phase A)				Inverter Terminal Voltage
	S_1	S_2	S_3	S_4	v_{AZ}
P	On	On	Off	Off	E
O	Off	On	On	Off	0
N	Off	Off	On	On	$-E$

current i_A, one of the two clamping diodes is turned on. For instance, a positive load current ($i_A > 0$) forces D_{Z1} to turn on, and the terminal A is connected to the neutral point Z through the conduction of D_{Z1} and S_2.

It can be observed from Table 8.2-1 that switches S_1 and S_3 operate in a complementary manner. With one switched on, the other must be off. Similarly, S_2 and S_4 are a complementary pair as well.

Figure 8.2-2 shows an example of switching state and gate signal arrangements, where v_{g1} to v_{g4} are the gate signals for S_1 to S_4, respectively. The gate signals can be generated by carrier-based modulation, space vector modulation, or selective harmonic elimination schemes. The waveform for v_{AZ} has three voltage levels, $+E$, 0, and $-E$, based on which the inverter is referred to as a three-level inverter.

Figure 8.2-3 shows how the line-to-line voltage waveform is obtained. The inverter terminal voltages v_{AZ}, v_{BZ}, and v_{CZ} are three-phase balanced with a phase shift of $2\pi/3$ between each other. The line-to-line voltage v_{AB} can be found from $v_{AB} = v_{AZ} - v_{BZ}$, which contains five voltage levels ($+2E$, $+E$, 0, $-E$, and $-2E$).

8.2.3 Commutation

To investigate the commutation of switching devices in the NPC inverter, consider a transition from switching state [O] to [P] by turning S_3 off and turning S_1 on. Figure 8.2-4a shows the gate signals v_{g1} to v_{g4} for switches S_1 to S_4, respectively. Similar to the gating arrangement in the two-level inverter, a blanking time of δ is required for the complementary switch pair S_1 and S_3.

Figures 8.2-4b and 8.2-4c show the circuit diagram of the inverter leg A during commutation, where each of the active switches has a parallel resistor for static voltage sharing. According to the direction of the phase A load current i_A, the following two cases are investigated.

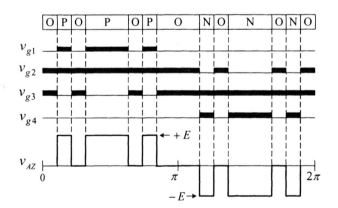

Figure 8.2-2 Switching states, gate signals and inverter terminal voltage v_{AZ}.

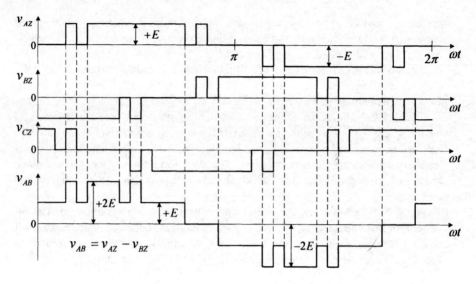

Figure 8.2-3 Inverter terminal and line-to-line voltage waveforms.

Case 1: Commutation With $i_A > 0$. The commutation process is illustrated in Fig. 8.2-4b. It is assumed that (a) the load current i_A is constant during the commutation due to the inductive load, (b) the dc bus capacitors C_{d1} and C_{d2} are sufficiently large such that the voltage across each capacitor is kept at E, and c) all the switches are ideal. In switching state [O], switches S_1 and S_4 are switched off while S_2 and S_3 conduct. The clamping diode D_{Z1} is turned on by the positive load current $(i_A > 0)$. The voltages across the on-state switches S_2 and S_3 are given by $v_{S2} = v_{S3} = 0$, while the voltage on each of the off-state switches S_1 and S_4 is equal to E.

During the δ interval, S_3 is being turned off. The paths of i_A remain unchanged. When S_3 is completely switched off, the voltages across S_3 and S_4 become $v_{S3} = v_{S4} = E/2$ due to the static voltage sharing resistors R_3 and R_4.

In switching state [P], the top switch S_1 is gated on ($v_{S1} = 0$). The clamping diode D_{Z1} is reverse-biased and thus turned off. The load current i_A is commutated from D_{Z1} to S_1. Since both S_3 and S_4 have already been in the off-state, the voltage across these two switches is equally divided by R_3 and R_4, leading to $v_{S3} = v_{S4} = E$.

Case 2: Commutation With $i_A < 0$. The commutation process with $i_A < 0$ is illustrated in Fig. 8.2-4c. In switching state [O], S_2 and S_3 conduct, and the clamping diode D_{Z2} is turned on by the negative load current i_A. The voltage across the off-state switches S_1 and S_4 is $v_{S1} = v_{S4} = E$.

During the δ interval, S_3 is being turned off. Since the inductive load current i_A cannot change its direction instantly, it forces diodes D_1 and D_2 to turn on, resulting in $v_{S1} = v_{S2} = 0$. The load current is commutated from S_3 to the diodes. During the S_3 turn-off transient, the voltage across S_4 will not be higher than E due to the clamping diode D_{Z2}, and it will also not be lower than E since the equivalent resistance of

8.2 Three-Level Inverter 147

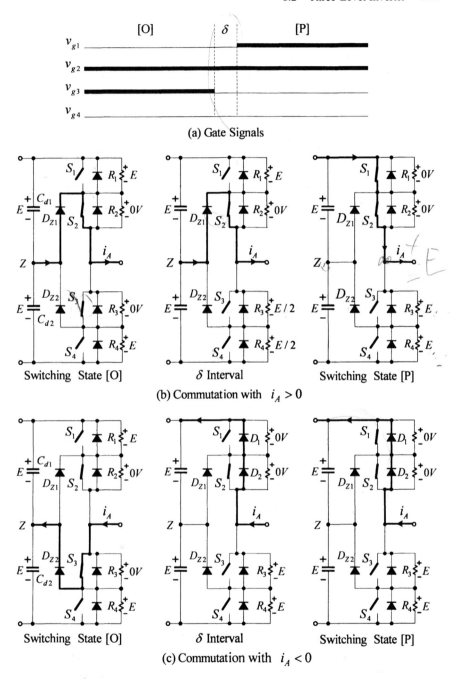

Figure 8.2-4 Commutation during a transition from switching state [O] to [P].

S_3 during turn-off is always lower than the off-state resistance of S_4. Therefore, v_{S3} increases from zero to E while v_{S4} is kept at E.

In switching state [P], the turn-on of S_1 does not affect the operation of the circuit. Although S_1 and S_2 are switched on, they do not carry the load current due to the conduction of D_1 and D_2.

It can be concluded that all the switching devices in the NPC inverter withstand only half of the dc bus voltage during the commutation from switching state [O] to [P]. Similarly, the same conclusion can be drawn for the commutation from [P] to [O], [N] to [O], or vice versa. Therefore, the switches in the NPC inverter do not have dynamic voltage sharing problem.

It should be pointed out that the switching between [P] and [N] is prohibited for two reasons: (a) It involves all four switches in an inverter leg, two being turned on and the other two being commutated off, during which the dynamic voltage on each switch may not be kept same; and (b) the switching loss is doubled.

It is worth noting that the static voltage sharing resistors R_1 to R_4 may be omitted if the leakage current of the top and bottom switches (S_1 and S_4) in each inverter leg is selected to be lower than that of the inner two switches (S_2 and S_3). In doing so, the voltages across the top and bottom switches, which tend to be higher than those of the inner switches, are clamped to E by the clamping diodes in steady state. As a result, the voltage on each of the inner two switches is also equal to E, and the static voltage equalization is achieved.

To summarize, the three-level NPC inverter offers the following features:

- **No dynamic voltage sharing problem.** Each of the switches in the NPC inverter withstands only half of the total dc voltage during commutation.
- **Static voltage equalization without using additional components.** The static voltage equalization can be achieved when the leakage current of the top and bottom switches in an inverter leg is selected to be lower than that of the inner switches.
- **Low THD and *dv/dt*.** The waveform of the line-to-line voltages is composed of five voltage levels, which leads to lower THD and *dv/dt* in comparison to the two-level inverter operating at the same voltage rating and device switching frequency.

However, the NPC inverter has some drawbacks such as additional clamping diodes, complicated PWM switching pattern design, and possible deviation of neutral point voltage.

8.3 SPACE VECTOR MODULATION

Various space vector modulation (SVM) schemes have been proposed for the three-level NPC inverter [4-7]. This section presents a "conventional" SVM scheme for the NPC inverter, followed by a modified SVM scheme for even-order harmonic elimination [8].

8.3.1 Stationary Space Vectors

As indicated earlier, the operation of each inverter phase leg can be represented by three switching states [P], [O], and [N]. Taking all three phases into account, the inverter has a total of 27 possible combinations of switching states. As listed in Table 8.3-1, these three-phase switching states are represented by three letters in square brackets for the inverter phases A, B, and C.

To find the relationship between the switching states and their corresponding space voltage vectors, we can follow the same procedures presented in Chapter 6. The 27 switching states listed in the table correspond to 19 voltage vectors whose space vector diagram is given in Fig. 8.3-1. Based on their magnitude (length), the voltage vectors can be divided into four groups:

- **Zero vector** ($\vec{V}_0$), representing three switching states [PPP], [OOO], and [NNN]. The magnitude of $\vec{V}_0$ is zero.
- **Small vectors** ($\vec{V}_1$ to $\vec{V}_6$), all having a magnitude of $V_d/3$. Each small vector has two switching states, one containing [P] and the other containing [N], and therefore can be further classified into a P- or N-type small vector.
- **Medium vectors** ($\vec{V}_7$ to $\vec{V}_{12}$), whose magnitude is $\sqrt{3}V_d/3$.
- **Large vectors** ($\vec{V}_{13}$ to $\vec{V}_{18}$), all having a magnitude of $2V_d/3$.

8.3.2 Dwell Time Calculation

To facilitate the dwell time calculation, the space vector diagram of Fig. 8.3-1 can be divided into six triangular **sectors** (I to VI), each of which can be further divided into four triangular **regions** (1 to 4) as illustrated in Fig. 8.3-2. The switching states of all the vectors are also shown in the figure.

Similar to the SVM algorithm for the two-level inverter, the space vector modulation for the NPC inverter is also based on "volt-second balancing" principle; that is, the product of the reference voltage $\vec{V}_{ref}$ and sampling period T_s equals the sum of the voltage multiplied by the time interval of chosen space vectors. In the NPC inverter, the reference vector $\vec{V}_{ref}$ can be synthesized by three nearest stationary vectors. For instance, when $\vec{V}_{ref}$ falls into region 2 of sector I as shown in Fig. 8.3-3, the three nearest vectors are $\vec{V}_1$, $\vec{V}_2$, and $\vec{V}_7$, from which

$$\vec{V}_1 T_a + \vec{V}_7 T_b + \vec{V}_2 T_c = \vec{V}_{ref} T_s$$
$$T_a + T_b + T_c = T_s$$
(8.3-1)

where T_a, T_b, and T_c are the dwell times for $\vec{V}_1$, $\vec{V}_7$, and $\vec{V}_2$, respectively. Note that $\vec{V}_{ref}$ can also be synthesized by other space vectors instead of the "nearest three." However, it will cause higher harmonic distortion in the inverter output voltage, which is undesirable in most cases.

Chapter 8 Diode-Clamped Multilevel Inverters

Table 8.3-1 Voltage Vectors and Switching States

Space Vector		Switching State		Vector Classification	Vector Magnitude
$\vec{V}_0$		[PPP][OOO] [NNN]		Zero vector	0
		P-type	**N-type**		
$\vec{V}_1$	$\vec{V}_{1P}$	[POO]		Small vector	$\frac{1}{3}V_d$
	$\vec{V}_{1N}$		[ONN]		
$\vec{V}_2$	$\vec{V}_{2P}$	[PPO]			
	$\vec{V}_{2N}$		[OON]		
$\vec{V}_3$	$\vec{V}_{3P}$	[OPO]			
	$\vec{V}_{3N}$		[NON]		
$\vec{V}_4$	$\vec{V}_{4P}$	[OPP]			
	$\vec{V}_{4N}$		[NOO]		
$\vec{V}_5$	$\vec{V}_{5P}$	[OOP]			
	$\vec{V}_{5N}$		[NNO]		
$\vec{V}_6$	$\vec{V}_{6P}$	[POP]			
	$\vec{V}_{6N}$		[ONO]		
$\vec{V}_7$		[PON]		Medium vector	$\frac{\sqrt{3}}{3}V_d$
$\vec{V}_8$		[OPN]			
$\vec{V}_9$		[NPO]			
$\vec{V}_{10}$		[NOP]			
$\vec{V}_{11}$		[ONP]			
$\vec{V}_{12}$		[PNO]			
$\vec{V}_{13}$		[PNN]		Large vector	$\frac{2}{3}V_d$
$\vec{V}_{14}$		[PPN]			
$\vec{V}_{15}$		[NPN]			
$\vec{V}_{16}$		[NPP]			
$\vec{V}_{17}$		[NNP]			
$\vec{V}_{18}$		[PNP]			

8.3 Space Vector Modulation 151

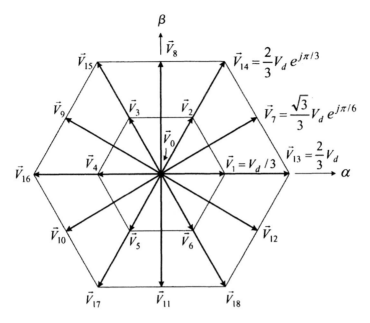

Figure 8.3-1 Space vector diagram of the NPC inverter.

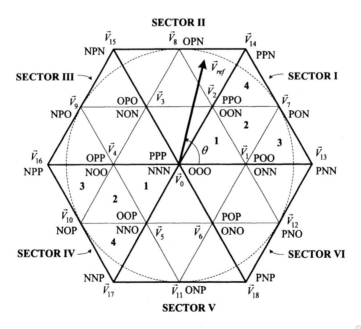

Figure 8.3-2 Division of sectors and regions.

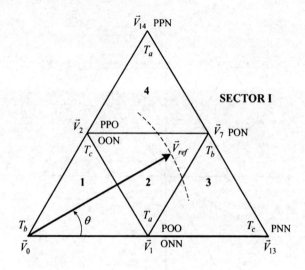

Figure 8.3-3 Voltage vectors and their dwell times.

The voltage vectors $\vec{V}_1, \vec{V}_2, \vec{V}_7,$ and $\vec{V}_{ref}$ in Fig. 8.3-3 can be expressed as

$$\vec{V}_1 = \frac{1}{3}V_d, \quad \vec{V}_2 = \frac{1}{3}V_d e^{j\pi/3}, \quad \vec{V}_7 = \frac{\sqrt{3}}{3}V_d e^{j\pi/6}, \quad \text{and} \quad \vec{V}_{ref} = V_{ref} e^{j\theta} \qquad (8.3\text{-}2)$$

Substituting (8.3-2) into (8.3-1) yields

$$\frac{1}{3}V_d T_a + \frac{\sqrt{3}}{3}V_d e^{j\pi/6} T_b + \frac{1}{3}V_d e^{j\pi/3} T_c = V_{ref} e^{j\theta} T_s \qquad (8.3\text{-}3)$$

from which

$$\frac{1}{3}V_d T_a + \frac{\sqrt{3}}{3}V_d \left(\cos\frac{\pi}{6} + j\sin\frac{\pi}{6}\right) T_b + \frac{1}{3}V_d \left(\cos\frac{\pi}{3} + j\sin\frac{\pi}{3}\right) T_c$$
$$= V_{ref}(\cos\theta + j\sin\theta) T_s \qquad (8.3\text{-}4)$$

Splitting (8.3-4) into the real and imaginary parts, we have

$$\text{Re:} \quad T_a + \frac{3}{2}T_b + \frac{1}{2}T_c = 3\frac{V_{ref}}{V_d}(\cos\theta) T_s$$
$$\text{Im:} \quad \frac{3}{2}T_b + \frac{\sqrt{3}}{2}T_c = 3\frac{V_{ref}}{V_d}(\sin\theta) T_s \qquad (8.3\text{-}5)$$

8.3 Space Vector Modulation

Solve (8.3-5) together with $T_s = T_a + T_b + T_c$ for dwell times

$$T_a = T_s[1 - 2m_a \sin \theta]$$

$$T_b = T_s\left[2m_a \sin\left(\frac{\pi}{3} + \theta\right) - 1\right] \quad \text{for } \leq \theta < \pi/3 \quad (8.3\text{-}6)$$

$$T_c = T_s\left[1 - 2m_a \sin\left(\frac{\pi}{3} - \theta\right)\right]$$

where m_a is the **modulation index**, defined by

$$m_a = \sqrt{3}\frac{V_{ref}}{V_d} \quad (8.3\text{-}7)$$

The maximum length of the reference vector $\vec{V}_{ref}$ corresponds to the radius of the largest circle that can be inscribed within the hexagon of Fig. 8.3-2, which happens to be the length of the medium voltage vectors

$$V_{ref,max} = \sqrt{3}V_d/3$$

Substituting $V_{ref,max}$ into (8.3-7) yields the maximum modulation index

$$m_{a,max} = \sqrt{3}\frac{V_{ref,max}}{V_d} = 1 \quad (8.3\text{-}8)$$

from which the range of m_a is

$$0 \leq m_a \leq 1 \quad (8.3\text{-}9)$$

Table 8.3-2 gives the equations for the calculation of dwell times for $\vec{V}_{ref}$ in sector I.

Table 8.3-2 Dwell Time Calculation for $\vec{V}_{ref}$ in Sector I

Region		T_a		T_b		T_c
1	$\vec{V}_1$	$T_s\left[2m_a \sin\left(\frac{\pi}{3} - \theta\right)\right]$	$\vec{V}_0$	$T_s\left[1 - 2m_a \sin\left(\frac{\pi}{3} + \theta\right)\right]$	$\vec{V}_2$	$T_s[2m_a \sin \theta]$
2	$\vec{V}_1$	$T_s[1 - 2m_a \sin \theta]$	$\vec{V}_7$	$T_s\left[2m_a \sin\left(\frac{\pi}{3} + \theta\right) - 1\right]$	$\vec{V}_2$	$T_s\left[1 - 2m_a \sin\left(\frac{\pi}{3} - \theta\right)\right]$
3	$\vec{V}_1$	$T_s\left[2 - 2m_a \sin\left(\frac{\pi}{3} + \theta\right)\right]$	$\vec{V}_7$	$T_s[2m_a \sin \theta]$	$\vec{V}_{13}$	$T_s\left[2m_a \sin\left(\frac{\pi}{3} - \theta\right) - 1\right]$
4	$\vec{V}_{14}$	$T_s[2m_a \sin \theta - 1]$	$\vec{V}_7$	$T_s\left[2m_a \sin\left(\frac{\pi}{3} - \theta\right)\right]$	$\vec{V}_2$	$T_s\left[2 - 2m_a \sin\left(\frac{\pi}{3} + \theta\right)\right]$

The equations in Table 8.3-2 can also be used to calculate the dwell times when $\vec{V}_{ref}$ is in other sectors (II to VI) provided that a multiple of $\pi/3$ is subtracted from the actual angular displacement θ such that the modified angle falls into the range between zero and $\pi/3$ for use in the equations. The reader can refer to Chapter 6 for details.

8.3.3 Relationship Between $\vec{V}_{ref}$ Location and Dwell Times

To demonstrate the relationship between the $\vec{V}_{ref}$ location and dwell times, consider an example shown in Fig. 8.3-4. Assuming that the head of $\vec{V}_{ref}$ points to the center Q of region 4, the dwell times for the nearest three vectors $\vec{V}_2$, $\vec{V}_7$, and $\vec{V}_{14}$ should be identical since the distance from Q to these vectors is the same. This can be verified by substituting $m_a = 0.882$ and $\theta = 49.1°$ into the equations in Table 8.3-2, from which the calculated dwell times are $T_a = T_b = T_c = 0.333 T_s$.

With $\vec{V}_{ref}$ moving toward $\vec{V}_2$ from Q along the dashed line, the influence of $\vec{V}_2$ on $\vec{V}_{ref}$ becomes stronger, which translates into a longer dwell time for $\vec{V}_2$. When $\vec{V}_{ref}$ is identical to $\vec{V}_2$, the dwell time T_c for $\vec{V}_2$ reaches its maximum value ($T_c = T_s$) while T_a and T_b for $\vec{V}_{14}$ and $\vec{V}_7$ diminish to zero.

8.3.4 Switching Sequence Design

The neutral point voltage v_Z, which is defined as the voltage between the neutral point Z and the negative dc bus, normally varies with the switching state of the NPC

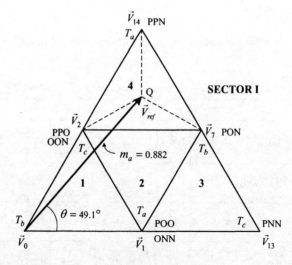

Figure 8.3-4 An example to demonstrate the relationship between the location of $\vec{V}_{ref}$ and dwell times.

inverter. When designing the switching sequence, we should minimize the effect of the switching state on neutral point voltage deviation. Taking into account the two requirements presented in Chapter 6 for the two-level inverter, the overall requirements for switching sequence design in the NPC inverter are as follows:

(a) The transition from one switching state to the next involves only two switches in the same inverter leg, one being switched on and the other switched off.
(b) The transition for $\vec{V}_{ref}$ moving from one sector (or region) to the next requires no or minimum number of switchings.
(c) The effect of switching state on the neutral-point voltage deviation is minimized.

(1) Effect of Switching States on Neutral-Point Voltage Deviation.
The effect of switching states on neutral voltage deviation is illustrated in Fig. 8.3-5. When the inverter operates with switching state [PPP] of zero vector $\vec{V}_0$,

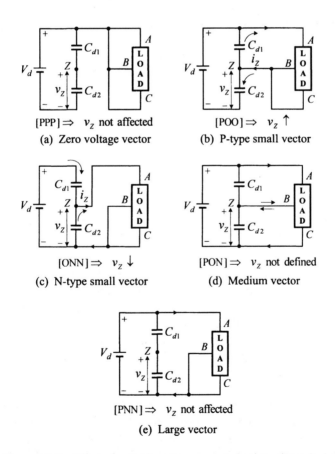

Figure 8.3-5 Effect of switching states on neutral point voltage deviation.

156 Chapter 8   Diode-Clamped Multilevel Inverters

the upper two switches in each of the three inverter legs are turned on, connecting the inverter terminals A, B, and C to the positive dc bus as shown in Fig. 8.3-5a. Since the neutral point Z is left unconnected, this switching state does not affect v_Z. Similarly, the other two zero switching states, [OOO] and [NNN], do not cause v_Z to shift either.

Figure 8.3-5b shows the inverter operation with P-type switching state [POO] of small vector $\vec{V}_1$. Since the three-phase load is connected between the positive dc bus and neutral point Z, the neutral current i_Z flows into Z, causing v_Z to increase. On the contrary, the N-type switching state [ONN] of $\vec{V}_1$ makes v_Z to decrease as shown in Fig. 8.3-5c.

The medium-voltage vectors also affect the neutral-point voltage. For medium vector $\vec{V}_7$ with switching state [PON] in Fig. 8.3-5d, load terminals A, B, and C are connected to the positive bus, the neutral point, and the negative bus, respectively. Depending on the inverter operating conditions, the neutral-point voltage v_Z may rise or drop.

Considering a large vector $\vec{V}_{13}$ with switching state [PNN] shown in Fig. 8.3-5e, the load terminals are connected between the positive and negative dc buses. The neutral point Z is left unconnected, and thus the neutral voltage is not affected.

It can be summarized that

- Zero vector $\vec{V}_0$ does not affect the neutral point voltage v_Z.
- Small vectors $\vec{V}_1$ to $\vec{V}_6$ have a dominant influence on v_Z. A P-type small vector makes v_Z rise, while an N-type small vector causes v_Z to decline.
- Medium vectors $\vec{V}_7$ to $\vec{V}_{12}$ also affect v_Z, but the direction of voltage deviation is undefined.
- The large vectors $\vec{V}_{13}$ to $\vec{V}_{18}$ do not play a role in neutral-point voltage deviation.

Note that the above summary is made under the assumption that the inverter is in normal (motoring) operating mode. The effect of the regenerative operating mode on neutral point voltage shift will be addressed later.

(2) Switching Sequence with Minimal Neutral-Point Voltage Deviation.

As mentioned earlier, a P-type small vector causes the neutral-point voltage v_Z to rise while an N-type small vector makes v_Z fall. To minimize the neutral-point voltage deviation, the dwell time of a given small vector can be equally distributed between the P- and N-type switching states over a sampling period. According to the triangular region that the reference vector $\vec{V}_{ref}$ lies in, the following two cases are investigated.

Case 1: One Small Vector Among Three Selected Vectors. When the reference vector $\vec{V}_{ref}$ is in region 3 or 4 of sector I shown in Fig. 8.3-3, only one of the three selected vectors is the small vector. Assuming that $\vec{V}_{ref}$ falls into region 4, it can be synthesized by $\vec{V}_2$, $\vec{V}_7$, and $\vec{V}_{14}$. The small vector $\vec{V}_2$ has two switching states [PPO] and [OON]. To minimize the neutral voltage deviation, the dwell time for $\vec{V}_2$ should be equally distributed between the P- and N-type states. Figure 8.3-6 shows a typical seven-segment switching sequence for the NPC inverter, from which we can observe that

- The dwell times for the seven segments add up to the sampling period of the PWM pattern ($T_s = T_a + T_b + T_c$).
- Design requirement (a) is satisfied. For instance, the transition from [OON] to [PON] is accomplished by turning S_1 on and switching S_3 off, which involves only two switches.
- The dwell time T_c for $\vec{V}_2$ is equally divided between the P- and N-type switching states, which satisfies design requirement (c).
- Among the four switching devices in an inverter leg, only two are tuned on and off once per sampling period. Assuming that the transition for $\vec{V}_{ref}$ moving from one sector (or region) to the next does not involve any switchings, the **device switching frequency** $f_{sw,dev}$ is equal to half of the sampling frequency f_{sp}, that is,

$$f_{sw,dev} = f_{sp}/2 = 1/(2T_s) \qquad (8.3\text{-}10)$$

Case 2: Two Small Vectors Among Three Selected Vectors. When $\vec{V}_{ref}$ is in region 1 or 2 of sector I in Fig. 8.3-3, two of the three selected vectors are small vectors. To reduce the neutral voltage deviation, each of the two regions is further divided into two subregions as shown in Fig. 8.3-7. Assuming that $\vec{V}_{ref}$ lies in region

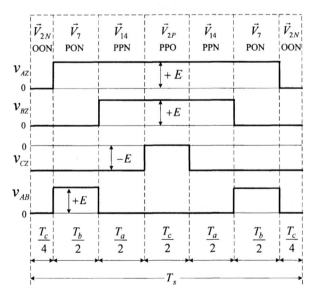

Figure 8.3-6 Seven-segment switching sequence for $\vec{V}_{ref}$ in sector I-4.

158   Chapter 8   Diode-Clamped Multilevel Inverters

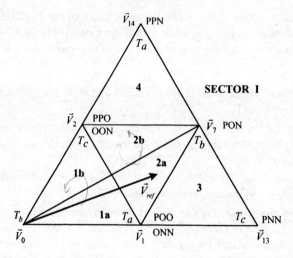

Figure 8.3-7 Division of six regions of sector I for the minimization of neutral point voltage deviation.

2a, it can be approximated by $\vec{V}_1$, $\vec{V}_2$, and $\vec{V}_7$. Since $\vec{V}_{ref}$ is closer to $\vec{V}_1$ than $\vec{V}_2$, the corresponding dwell time T_a for $\vec{V}_1$ is longer than T_c for $\vec{V}_2$. The vector $\vec{V}_1$ is referred to as **dominant small vector**, whose dwell time is equally divided between $\vec{V}_{1P}$ and $\vec{V}_{1N}$ as shown in Table 8.3-3.

Based on the above discussions, all the switching sequences in sectors I and II are summarized in Table 8.3-4. It can be observed that (1) when $\vec{V}_{ref}$ crosses the border between sectors I and II, the transition does not involve any switchings; and (2) an extra switching takes place when $\vec{V}_{ref}$ moves from region a to b within a sector. The graphical representation is illustrated in Fig. 8.3-8, where the large and small circles are the steady-state trajectories of $\vec{V}_{ref}$ and the dots represent the locations at which an extra switching takes place. Since each of these extra switchings involves only two devices (out of twelve) and there are only six extra switchings per cycle of fundamental frequency, the average switching frequency of the device is increased to

$$f_{sw,dev} = f_{sp}/2 + f_1/2 \qquad (8.3\text{-}11)$$

Table 8.3-3 Seven-Segment Switching Sequence for $\vec{V}_{ref}$ in Sector I-2a

Segment:	1st	2nd	3rd	4th	5th	6th	7th
Voltage Vector:	$\vec{V}_{1N}$	$\vec{V}_{2N}$	$\vec{V}_7$	$\vec{V}_{1P}$	$\vec{V}_7$	$\vec{V}_{2N}$	$\vec{V}_{1N}$
Switching State:	[ONN]	[OON]	[PON]	[POO]	[PON]	[OON]	[ONN]
Dwell Time:	$\dfrac{T_a}{4}$	$\dfrac{T_c}{2}$	$\dfrac{T_b}{2}$	$\dfrac{T_a}{2}$	$\dfrac{T_b}{2}$	$\dfrac{T_c}{2}$	$\dfrac{T_a}{4}$

8.3 Space Vector Modulation

Table 8.3-4 Seven-Segment Switching Sequence

Sector I														
Sgmt	1a		1b		2a		2b		3		4			
1st	$\vec{V}_{1N}$	[ONN]	$\vec{V}_{2N}$	[OON]	$\vec{V}_{1N}$	[ONN]	$\vec{V}_{2N}$	[OON]	$\vec{V}_{1N}$	[ONN]	$\vec{V}_{2N}$	[OON]		
2nd	$\vec{V}_{2N}$	[OON]	$\vec{V}_0$	[OOO]	$\vec{V}_{2N}$	[OON]	$\vec{V}_7$	[PON]	$\vec{V}_{13}$	[PNN]	$\vec{V}_7$	[PON]		
3rd	$\vec{V}_0$	[OOO]	$\vec{V}_{1P}$	[POO]	$\vec{V}_7$	[PON]	$\vec{V}_{1P}$	[POO]	$\vec{V}_7$	[PON]	$\vec{V}_{14}$	[PPN]		
4th	$\vec{V}_{1P}$	[POO]	$\vec{V}_{2P}$	[PPO]	$\vec{V}_{1P}$	[POO]	$\vec{V}_{2P}$	[PPO]	$\vec{V}_{1P}$	[POO]	$\vec{V}_{2P}$	[PPO]		
5th	$\vec{V}_0$	[OOO]	$\vec{V}_{1P}$	[POO]	$\vec{V}_7$	[PON]	$\vec{V}_{1P}$	[POO]	$\vec{V}_7$	[PON]	$\vec{V}_{14}$	[PPN]		
6th	$\vec{V}_{2N}$	[OON]	$\vec{V}_0$	[OOO]	$\vec{V}_{2N}$	[OON]	$\vec{V}_7$	[PON]	$\vec{V}_{13}$	[PNN]	$\vec{V}_7$	[PON]		
7th	$\vec{V}_{1N}$	[ONN]	$\vec{V}_{2N}$	[OON]	$\vec{V}_{1N}$	[ONN]	$\vec{V}_{2N}$	[OON]	$\vec{V}_{1N}$	[ONN]	$\vec{V}_{2N}$	[OON]		
Sector II														
Sgmt	1a		1b		2a		2b		3		4			
1st	$\vec{V}_{2N}$	[OON]	$\vec{V}_{3N}$	[NON]	$\vec{V}_{2N}$	[OON]	$\vec{V}_{3N}$	[NON]	$\vec{V}_{2N}$	[OON]	$\vec{V}_{3N}$	[NON]		
2nd	$\vec{V}_0$	[OOO]	$\vec{V}_{2N}$	[OON]	$\vec{V}_8$	[OPN]	$\vec{V}_{2N}$	[OON]	$\vec{V}_8$	[OPN]	$\vec{V}_{15}$	[NPN]		
3rd	$\vec{V}_{3P}$	[OPO]	$\vec{V}_0$	[OOO]	$\vec{V}_{3P}$	[OPO]	$\vec{V}_8$	[OPN]	$\vec{V}_{14}$	[PPN]	$\vec{V}_8$	[OPN]		
4th	$\vec{V}_{2P}$	[PPO]	$\vec{V}_{3P}$	[OPO]	$\vec{V}_{2P}$	[PPO]	$\vec{V}_{3P}$	[OPO]	$\vec{V}_{2P}$	[PPO]	$\vec{V}_{3P}$	[OPO]		
5th	$\vec{V}_{3P}$	[OPO]	$\vec{V}_0$	[OOO]	$\vec{V}_{3P}$	[OPO]	$\vec{V}_8$	[OPN]	$\vec{V}_{14}$	[PPN]	$\vec{V}_8$	[OPN]		
6th	$\vec{V}_0$	[OOO]	$\vec{V}_{2N}$	[OON]	$\vec{V}_8$	[OPN]	$\vec{V}_{2N}$	[OON]	$\vec{V}_8$	[OPN]	$\vec{V}_{15}$	[NPN]		
7th	$\vec{V}_{2N}$	[OON]	$\vec{V}_{3N}$	[NON]	$\vec{V}_{2N}$	[OON]	$\vec{V}_{3N}$	[NON]	$\vec{V}_{2N}$	[OON]	$\vec{V}_{3N}$	[NON]		

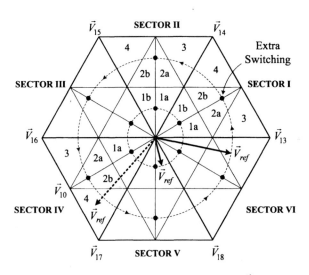

Figure 8.3-8 Graphical representation of extra switchings when $\vec{V}_{ref}$ moves from region a to b.

8.3.5 Inverter Output Waveforms and Harmonic Content

Figure 8.3-9 shows the simulated waveforms for the NPC inverter operating at $f_1 = 60$ Hz, $T_s = 1/1080$ s, $f_{sw,dev} = 1080/2 + 60/2 = 570$ Hz, and $m_a = 0.8$. The inverter is loaded with a three-phase inductive load with a power factor of 0.9. The gate signals v_{g1} and v_{g4} are for the switches S_1 and S_4 of the inverter circuit in Fig. 8.2-1. Since the inner switches S_2 and S_3 operate complementarily with S_4 and S_1, their gatings are not shown.

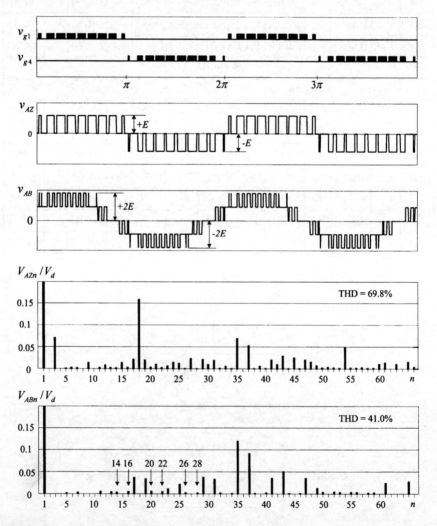

Figure 8.3-9 Simulated voltage waveforms of the NPC inverter ($f_1 = 60$ Hz, $T_s = 1/1080$ s, $f_{sw,dev} = 570$ Hz, and $m_a = 0.8$).

The waveform of the inverter terminal voltage v_{AZ} is composed of three voltage levels, while the inverter line-to-line voltage v_{AB} has five voltage levels. The waveform for v_{AZ} contains triplen harmonics with the 3rd and 18th being dominant. Since the triplen harmonics are of zero sequence, they do not appear in the line-to-line voltage v_{AB}. However, v_{AB} contains even-order harmonics such as the 14th and 16th in addition to odd-order harmonics. This is due to the fact that waveform of v_{AB} produced by the SVM scheme is not half-wave symmetrical.

The dominant harmonics in v_{AB} are the 17th and 19th, centered around the 18th harmonic whose frequency is 1080 Hz. As discussed in Chapter 6, this frequency can be considered as the equivalent **inverter switching frequency** $f_{sw,inv}$, which is approximately twice the device switching frequency $f_{sw,dev}$.

The harmonic content and THD of v_{AB} versus m_a are illustrated in Fig. 8.3-10, where V_{ABn} is the rms value of the nth-order harmonic voltage. The waveform of v_{AB} contains all the low-order harmonics except for triplen harmonics. The magni-

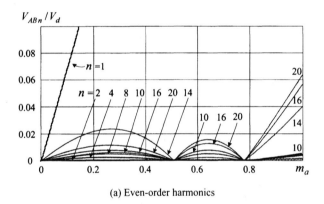

(a) Even-order harmonics

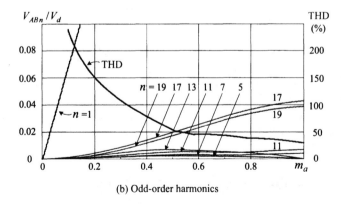

(b) Odd-order harmonics

Figure 8.3-10 Harmonic content and THD of the inverter line-to-line voltage v_{AB} ($f_1 = 60$ Hz, $T_s = 1/1080$ s, and $f_{sw,dev} = 570$ Hz).

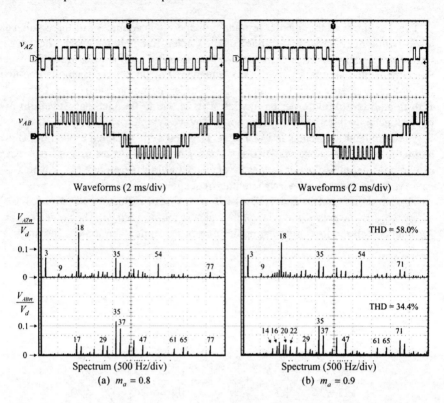

Figure 8.3-11 Measured waveforms and their harmonic spectra (f_1 = 60 Hz, T_s = 1/1080 s, and $f_{sw,dev}$ = 570 Hz).

tude of most even-order harmonics peaks at m_a = 1. The maximum rms fundamental voltage occurs at m_a = 1, at which

$$V_{AB1,\max} = 0.707 V_d \qquad (8.3\text{-}12)$$

Fig. 8.3-11 shows waveforms measured from a laboratory three-level NPC inverter operating at f_1 = 60 Hz, T_s = 1/1080 s, and $f_{sw,dev}$ = 570 Hz with the modulation index m_a equal to 0.8 and 0.9, respectively. The measured waveforms and their spectrum at m_a = 0.8 correlate closely with the simulated results in Fig. 8.3-9. It can also be observed that the magnitude of the even-order harmonics at m_a = 0.9 is much higher than that for m_a = 0.8, which is consistent with the harmonic content illustrated in Fig. 8.3-10.

8.3.6 Even-Order Harmonic Elimination

The mechanism of even-order harmonic generation and the reasons for its elimination have been discussed in Chapter 6 for the two-level inverter. They can be equally applied to the three-level NPC inverter and therefore are not repeated here.

Figure 8.3-12 shows two valid switching sequences for $\vec{V}_{ref}$ in sector IV-4 of the space vector diagram in Fig. 8.3-8. It can be observed that type-A sequence starts with an N-type small vector while type-B sequence commences with a P-type small vector. Although the waveforms of v_{AZ}, v_{BZ}, v_{CZ}, and v_{AB} in parts (a) and (b) seem quite different, they are essentially the same except for a small amount of time delay ($T_s/2$), which can be clearly observed if these waveforms are drawn for two or more consecutive sampling periods.

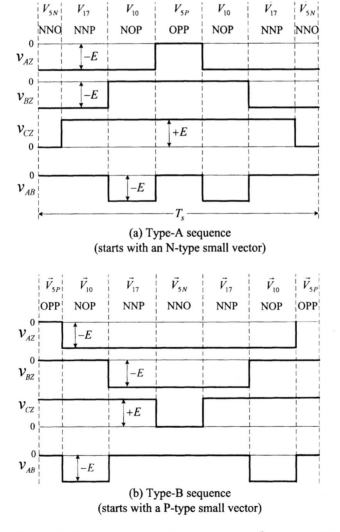

(a) Type-A sequence
(starts with an N-type small vector)

(b) Type-B sequence
(starts with a P-type small vector)

Figure 8.3-12 Two valid switching sequences for $\vec{V}_{ref}$ in sector IV-4.

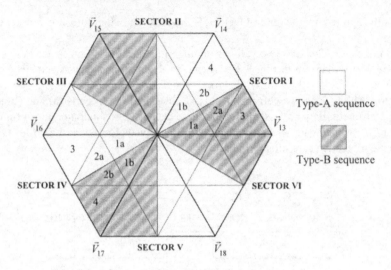

Figure 8.3-13 Alternative use of type-A and type-B switching sequences for even-order harmonic elimination.

In the conventional SVM scheme for the NPC inverter, only the type-A switching sequence is employed. To eliminate the even-order harmonics in v_{AB}, type-A and type-B switching sequences can be alternatively used as illustrated in Fig. 8.3-13. The reader can refer to Chapter 6 for the principle of even-order harmonic elimination. A complete set of the switching sequences for the modified SVM scheme is given in the appendix of this chapter. Compared with the conventional SVM, the modified scheme causes a slight increase in the device switching frequency. The amount of increase is given by $\Delta f_{sw} = f_1/2$, from which

$$f_{sw,dev} = f_{sp}/2 + f_1 \qquad (8.3\text{-}13)$$

Figure 8.3-14 shows waveforms measured from a laboratory NPC inverter with the modified SVM scheme. The inverter output voltage waveforms of v_{AZ} and v_{AB} are of half-wave symmetry, leading to the elimination of even-order harmonics. It is interesting to note that although the harmonic spectrum of v_{AB} differs from that in Fig. 8.3-11, its THD essentially remains unchanged.

8.4 NEUTRAL-POINT VOLTAGE CONTROL

As indicated earlier, the neutral-point voltage v_Z varies with the operating condition of the NPC inverter. If the neutral-point voltage deviates too far, an uneven voltage distribution takes place, which may lead to premature failure of the switching devices and cause an increase in the THD of the inverter output voltage.

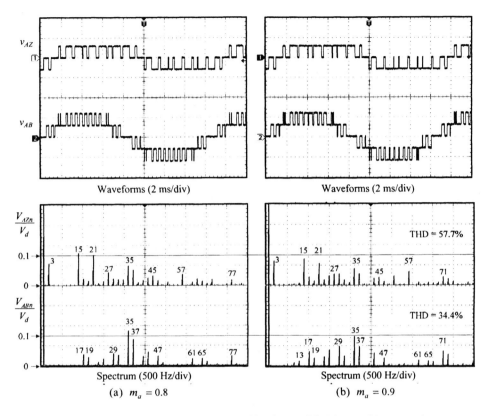

Figure 8.3-14 Measured waveforms produced by the modified SVM with even-order harmonic elimination (f_1 = 60 Hz, T_s = 1/1080 s, and $f_{sw,dev}$ = 600 Hz).

8.4.1 Causes of Neutral-Point Voltage Deviation

In addition to the influence of small- and medium-voltage vectors, the neutral-point voltage may also be affected by a number of other factors, including

- Unbalanced dc capacitors due to manufacturing tolerances
- Inconsistency in switching device characteristics
- Unbalanced three-phase operation

To minimize the neutral-point voltage shift, a feedback control scheme can be implemented, where the neutral-point voltage is detected and then controlled [7, 9, 10].

8.4.2 Effect of Motoring and Regenerative Operation

When the NPC inverter is used in the MV drives, the operating mode of the drive may also influence the neutral-point voltage. Figure 8.4-1 shows the effect of mo-

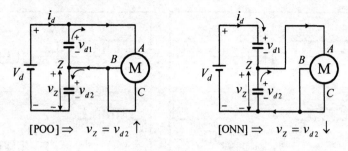

(a) Motoring operation

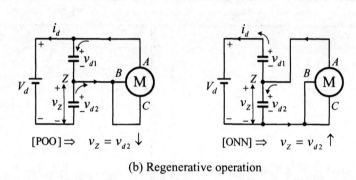

(b) Regenerative operation

Figure 8.4-1 Effect of drive operating modes on neutral-point voltage deviation.

toring and regenerative operations of the drive on neutral-point voltage shift. When the drive is in the motoring mode as shown in Fig. 8.4-1a where the dc current i_d flows from the dc source to the inverter, the P-type state [POO] of small vector $\vec{V}_1$ causes the neutral-point voltage v_Z to rise while the N-type state [ONN] makes v_Z to decline. An opposite action takes place in the regenerative mode in which the dc current reverses its direction as shown in Fig. 8.4-1b. This phenomenon should be taken into account when designing the feedback control for v_Z.

8.4.3 Feedback Control of Neutral-Point Voltage

The neutral-point voltage v_Z can be controlled by adjusting the time distribution between the P- and N-type states of a small-voltage vector. There always exists a small-voltage vector in each switching sequence, whose dwell time is divided into two subperiods, one for its P-type and the other for its N-type switching state. For instance, the dwell time T_a for $\vec{V}_{1P}$ and $\vec{V}_{1N}$, which is 50/50 split in Table 8.3-3, can be redistributed as

$$T_a = T_{aP} + T_{aN} \qquad (8.4\text{-}1)$$

where T_{aP} and T_{aN} are given by

$$T_{aP} = \frac{T_a}{2}(1 + \Delta t)$$
$$T_{aN} = \frac{T_a}{2}(1 - \Delta t)$$
for $-1 \leq \Delta t \leq 1$ \hfill (8.4-2)

The deviation of the neutral point voltage can be minimized by adjusting the incremental time interval Δt in (8.4-2) according to the detected dc capacitor voltages v_{d1} and v_{d2}. For instance, if $(v_{d1} - v_{d2})$ is greater than the maximum allowed dc voltage deviation ΔV_d for some reasons, we can increase T_{aP} and decrease T_{aN} by Δt ($\Delta t > 0$) simultaneously for the drive in a motoring mode. A reverse action ($\Delta t < 0$) should be taken when the drive is in a regenerative mode. The relationship between the capacitor voltages and the incremental time interval Δt is summarized in Table 8.4-1.

Figure 8.4-2 shows the simulated waveforms of the two dc capacitor voltages v_{d1} and v_{d2} with an equal initial value of 2800 V. To make v_{d1} and v_{d2} unbalanced on purpose, a resistor is connected in parallel with the bottom dc capacitor. When the NPC inverter operates at $t = 0$ with $f_1 = 60$ Hz, $T_s = 1/1080$ s, $m_a = 0.8$, and $V_d = 5600$ V, the voltage v_{d2} across the bottom capacitor starts to drop due to the discharging through the resistor while v_{d1} of the top capacitor starts to rise. At $t = 0.026$ s, the neutral point voltage control is activated, making dc capacitor voltages balanced at $t = 0.04$ s.

8.5 OTHER SPACE VECTOR MODULATION ALGORITHMS

In addition to the SVM schemes presented in the previous section, other SVM algorithms have been proposed for the NPC inverter [11–14]. Two of them are briefly introduced here.

8.5.1 Discontinuous Space Vector Modulation

The principle of the discontinuous (five-segment) SVM scheme presented in Chapter 6 for the two-level inverter can also be applied to the NPC inverter. The five-

Table 8.4-1 Relationship Between Capacitor Voltages and Incremental Time Interval Δt

Neutral Point Deviation Level	Motoring Mode $i_d > 0$	Regenerating Mode $i_d < 0$
$(v_{d1} - v_{d2}) > \Delta V_d$	$\Delta t > 0$	$\Delta t < 0$
$(v_{d2} - v_{d1}) > \Delta V_d$	$\Delta t < 0$	$\Delta t > 0$
$|v_{d1} - v_{d2}| < \Delta V_d$	$\Delta t = 0$	$\Delta t = 0$

ΔV_d – maximum allowed dc voltage deviation ($\Delta V_d > 0$).

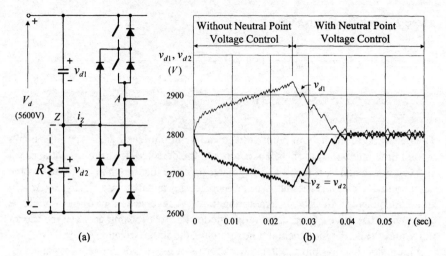

Figure 8.4-2 Neutral-point voltage control.

segment switching sequence can be arranged such that switching for the devices in one of the three inverter legs is avoided for a period of $\pi/3$ during the positive half-cycle and another $\pi/3$ during the negative half-cycle of the fundamental frequency. If the $\pi/3$ no-switching period is centered on the positive or negative peaks of the load current, the switching loss can be reduced. The reader can refer to references [11] and [12] for the details.

8.5.2 SVM Based on Two-Level Algorithm

The space vector diagram for the NPC inverter has an **outer hexagon** containing all 24 triangular regions and an **inner hexagon** with six triangular regions as shown in Fig. 8.5-1. The space vector diagram can be decomposed into six **small hexagons**, each of which centers at the six apexes of the inner hexagon [13]. Each of these small hexagons is composed of six triangular regions.

The position of the reference vector $\vec{V}_{ref}$ in the NPC space vector diagram determines which of the six small hexagons is selected. The selected hexagon is then shifted toward the center of the inner hexagon for the dwell time calculation and switching sequence design. Accordingly, $\vec{V}_{ref}$ should be also referred to the new co-ordinate system. In doing so, the SVM algorithm for the NPC inverter is simplified and can be performed in the same manner as for the two-level inverter.

8.6 HIGH-LEVEL DIODE-CLAMPED INVERTERS

To increase the inverter voltage rating and improve its waveform quality, high-level diode-clamped inverters can be employed. This section presents four- and five-level diode-clamped inverters.

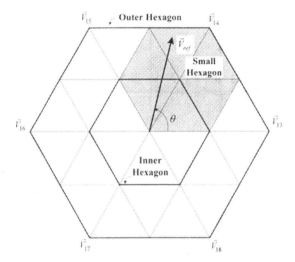

Figure 8.5-1 Space vector diagram of the three-level NPC inverter using two-level SVM algorithm.

8.6.1 Four- and Five-Level Diode-Clamped Inverters

Figure 8.6-1a shows the simplified per-phase diagram of the four-level diode clamped inverter. The inverter is composed of six active switches and a number of clamping diodes per phase. The dc capacitors C_d are shared by all three phases. It is assumed in the following analysis that the voltage across each capacitor is E and the total dc voltage V_d is equally divided by the capacitors ($V_d = 3E$).

The switch operating status and the inverter terminal voltage v_{AN} of the four-level inverter are summarized in Table 8.6-1, where '1' signifies that an active switch is turned on while "0" indicates that the switch is off. When the top three switches in leg A are on ($S_1 = S_2 = S_3 =$ "1"), v_{AN} is $3E$ whereas the conduction of the bottom three switches makes v_{AN} to be zero. When the inverter terminal A is connected to node X or Y of the capacitor circuit through the conduction of the middle three switches and clamping diodes, v_{AN} will be equal to $2E$ or E. Clearly, the waveform of v_{AN} is composed of four voltage levels: $3E$, $2E$, E, and 0. It can also be observed from the table that (a) in the four-level inverter, three switches conduct at any time instant and (b) switch pairs (S_1, S_1'), (S_2, S_2'), and (S_3, S_3') operate in a complementary manner.

It should be pointed out that the clamping diodes may withstand different reverse blocking voltages. For instance, when the inverter operates with $S_1 = S_2 = S_3 =$ "1", the anode of the clamping diodes D_1 and D_2 in Fig. 8.6-1a is connected to the positive dc bus. The voltage applied to D_1 and D_2 is then E and $2E$, respectively. In practice, the voltage rating for all the clamping diodes is normally selected to be the same as the active switches. As a result, two diodes should be in series for D_2 (denoted by $D_2 \times 2$ in the figure).

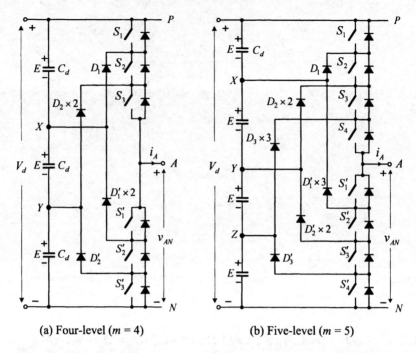

Figure 8.6-1 Per-phase diagram of four- and five-level diode-clamped inverters.

The per-phase circuit diagram of the five-level diode clamped inverter is shown in Fig. 8.6-1b, and the relationship between the switch status and v_{AN} for the five-level inverter is also given in Table 8.6-1. With various combinations of switch operating status, the waveform of v_{AN} contains five voltage levels: $4E$, $3E$, $2E$, E, and 0.

Table 8.6-2 lists the component count for the multilevel diode clamped inverters. Assuming that all the active switches and clamping diodes have the same voltage rating, the rated inverter output voltage is proportional to the number of active switches. This suggests that if the number of the switches is doubled, the maximum inverter output voltage increases twofold, and so does its output power. However, the number of clamping diodes increases dramatically with the voltage level. For example, the three-level inverter requires only six clamping diodes while the five-level inverter needs 36 clamping diodes. This is, in fact, one of the main reasons why the four- and five-level inverters are seldom found in industrial applications.

8.6.2 Carrier-Based PWM

The carrier-based modulation schemes presented in Chapter 7 for cascaded H-bridge multilevel inverters can also be used for the diode-clamped inverters. Figure

8.6 High-Level Diode-Clamped Inverters

Table 8.6-1 Switch Status and Inverter Terminal Voltage v_{AN}

\multicolumn{7}{c}{Switch Status}	
\multicolumn{7}{c}{Four-Level Inverter}	
S_1	S_2	S_3	S_1'	S_2'	S_3'	v_{AN}
1	1	1	0	0	0	$3E$
0	1	1	1	0	0	$2E$
0	0	1	1	1	0	E
0	0	0	1	1	1	0
\multicolumn{8}{c}{Five-Level Inverter}
S_1	S_2	S_3	S_4	S_1'	S_2'	S_3'	S_4'	v_{AN}
1	1	1	1	0	0	0	0	$4E$
0	1	1	1	1	0	0	0	$3E$
0	0	1	1	1	1	0	0	$2E$
0	0	0	1	1	1	1	0	E
0	0	0	0	1	1	1	1	0

8.6-2 illustrates the simulated waveforms of a four-level inverter modulated by an in-phase disposition (IPD) modulation scheme. The four-level inverter requires three carriers v_{cr1}, v_{cr2}, and v_{cr3}, which are disposed vertically, but all in phase. The amplitude modulation index m_a is equal to 0.9, and frequency modulation index m_f is 15.

The gate signals v_{g1}, v_{g2}, and v_{g3} for the top three switches S_1, S_2, and S_3 in Fig. 8.6-1a are generated at the intersections of the carrier waves and phase A modulation wave v_{mA}, respectively. The gatings for the bottom three devices S_1', S_2', and S_3' are complementary to v_{g1}, v_{g2}, and v_{g3} and therefore are not shown.

The inverter operates at $f_1 = 60$ Hz and $f_{sw,dev} = 300$ Hz, and it feeds a three-phase inductive load with a power factor of 0.9. The inverter terminal voltage v_{AN} has four voltage levels, and its line-to-line voltage v_{AB} contains seven voltage levels. The

Table 8.6-2 Component Count of Diode-Clamped Multilevel Inverters

Voltage Level	Active Switches	Clamping Diodes[a]	dc Capacitors
m	$6(m-1)$	$3(m-1)(m-2)$	$(m-1)$
3	12	6	2
4	18	18	3
5	24	36	4
6	30	60	5

[a]All diodes and active switches have the same voltage rating.

172 Chapter 8 Diode-Clamped Multilevel Inverters

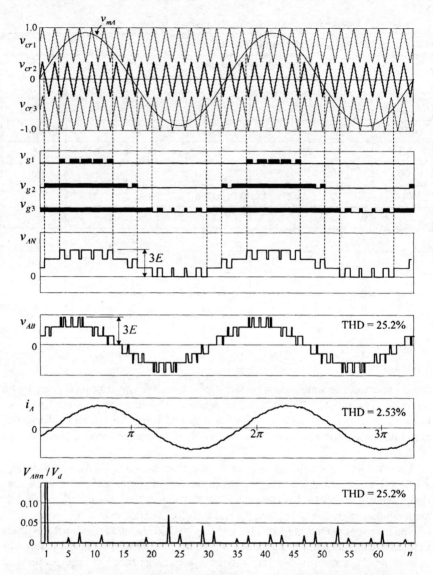

Figure 8.6-2 Simulated waveforms in the four-level inverter using IPD modulation (f_1 = 60 Hz, $f_{sw,dev}$ = 300 Hz, m_a = 0.9, and m_f = 15).

load current i_A is close to a sinusoid having a THD of only 2.53%. The waveform of v_{AB} contains low-order harmonics, such as the 5th and 7th, but their magnitudes are relatively low.

The harmonic content of v_{AB} is shown in Fig. 8.6-3. Although the four-level inverter with the IPD scheme generates low-order harmonics, the overall harmonic profile is quite good. At m_a = 1, the rms fundamental line-to-line voltage is V_{AB1} =

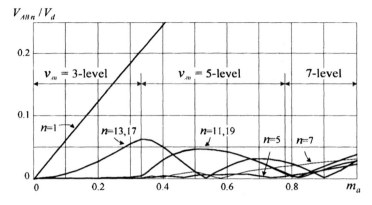

Figure 8.6-3 Harmonic content of v_{AB} in the four-level inverter (f_1 = 60 Hz, $f_{sw,dev}$ = 300 Hz, and m_f = 15).

$0.612V_d$, which can be further boosted by 15.5% to $0.707V_d$ using the third harmonic injection technique presented in Chapter 6. It is worth mentioning that v_{AB} is composed of three, five, and seven voltage levels for $0 \le m_a < 0.33$, $0.33 \le m_a < 0.74$, and $0.74 \le m_a \le 1.0$, respectively.

Figure 8.6-4 shows the waveforms of the four-level inverter modulated by an alternative phase opposite disposition (APOD) scheme, where all carriers are alternatively in opposite disposition. The inverter operates under the same conditions as those in the previous case. The THD of v_{AB} and i_A is 37.3% and 4.85%, respectively, much higher than those generated by the IPD modulation. The waveform of v_{AB} contains two pairs of dominant harmonics, (11th, 13th) and (17th, 19th), with relatively high magnitudes as shown in Figs. 8.6-4 and 8.6-5, but the 5th and 7th harmonics are eliminated.

In summary, the IPD modulation produces better harmonic profile than the APOD modulation, which is consistent with the conclusion made in Chapter 7. It should be noted that the phase-shifted modulation schemes cannot be utilized for the diode-clamped multilevel inverters.

8.7 SUMMARY

This chapter provides a comprehensive analysis on the three-level diode clamped inverter, also known as neutral-point clamped (NPC) inverter. A number of issues are investigated, including the inverter configuration, operating principle, space vector modulation (SVM) techniques, and neutral point voltage control. The emphasis of the chapter is on the SVM schemes, where the conventional SVM algorithm and its modified version for even-order harmonic elimination are discussed in detail. The harmonic profile and THD of the inverter output voltage are evaluated. Important concepts are illustrated with simulations and experiments.

174 Chapter 8 Diode-Clamped Multilevel Inverters

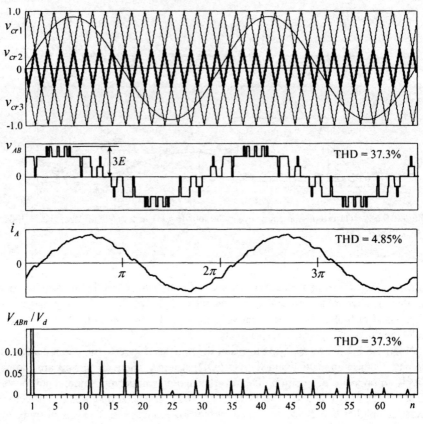

Figure 8.6-4 Simulated waveforms in the four-level inverter using APOD modulation (f_1 = 60 Hz, $f_{sw,dev}$ = 300 Hz, m_a = 0.9, and m_f = 15).

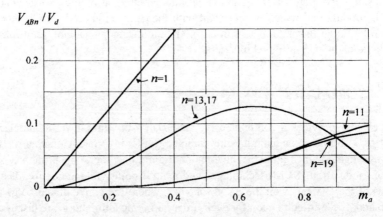

Figure 8.6-5 Harmonic content of v_{AB} in the four-level inverter (f_1 = 60 Hz, $f_{sw,dev}$ = 300 Hz, and m_f = 15).

In addition to the three-level inverter, four- and five-level diode-clamped inverters are also introduced. These inverters are seldom employed in practice mainly due to the increased number of clamping diodes and difficulties in dc capacitor voltage balance control.

REFERENCES

1. J. Sen and N. Butterworth, Analysis and Design of a Three-Phase PWM Converter System for Railway Traction Applications, *IEE Proceedings on Electric Power Applications*, Vol. 144, No. 5, pp. 357–371, 1997.
2. J. K. Steinke, H. Prenner, et al., New Variable Speed Drive with Proven Motor Friendly Performance for Medium Voltage Motors, *IEEE IEMD*, pp. 235–239, 1999.
3. J. P. Lyons, V. Vlatkovic, et al., Innovation IGCT Main Drives, *IEEE Industry Application Society Conference (IAS)*, pp. 2655–2661, 1999.
4. Y. H. Lee, B. S. Suh, et al., A Novel PWM Scheme for a Three Level Voltage Source Inverter with GTO Thyristors, *IEEE Transactions on Industry applications*, Vol. 32, No. 2, pp. 260–268, 1996.
5. R. Rojas, T. Ohnishi, et al., An Improved Voltage Vector Control Method for Neutral-Point-clamped Inverters, *IEEE Transactions on Power Electronics*, Vol. 10, No. 6, pp. 666–672, 1995.
6. Y. Shrivastava, C. K. Lee, et al., Comparison of RPWM and PWM Space Vector Switching Schemes for 3-Level Power Inverters, *IEEE Power Electronics Specialist Conference*, pp. 138–145, 2001.
7. D. Zhou, A Self-Balancing Space Vector Switching Modulator for Three-Level Motor Drives, *IEEE Power Electronics Specialist Conference (PESC)*, pp. 1369–1374, 2001.
8. D. W. Feng, B. Wu, et al., Space Vector Modulation for Neutral Point Clamped Multilevel Inverter with Even Order Harmonic Elimination, *Canadian Conference on Electrical and Computer Engineering (CCECE)*, pp. 1471–1475, 2004.
9. K. R. M. N. Ratnayake, et al., Novel PWM Scheme to Control Neutral Point Voltage Variation in Three-Lever Voltage Source Inverter, *IEEE Industry Application Society Conference (IAS)*, pp. 1950–1955, 1999.
10. D. Zhou and D. G. Rouaud, Experimental Comparisons of Space Vector Neutral Point Balancing Strategies for Three-Level Topology, *IEEE Transactions on Power Electronics*, Vol. 16, No. 6, pp. 872–879, 2001.
11. L. Helle, S. M. Nielsen, et al., Generalized Discontinuous DC-Link Balancing Modulation Strategy for Three-Level Inverters, *IEEE Power Conversion Conference*, pp. 359–366, 2002.
12. H. Kim, D. Jung, et al., A New Discontinuous PWM Strategy of Neutral Point Clamped Inverter, *IEEE Industry Application Society Conference (IAS)*, pp. 2017–2023, 2000.
13. J. H. Seo, C. H. Choi, et al., A New Simplified Space Vector PWM Method for Three-Level Inverters, *IEEE Transactions on Power Electronics*, Vol. 16, No. 4, pp. 545–555, 2001.
14. C. K. Lee, S. Y. R. Hui, et al., A Randomized Voltage Vector Switching Scheme for Three Level Power Inverters, *IEEE Transactions on Power Electronics*, Vol. 17, No. 1, pp. 94–100, 2002.

APPENDIX

SEVEN-SEGMENT SWITCHING SEQUENCE FOR THE THREE-LEVEL NPC INVERTER WITH EVEN-ORDER HARMONIC ELIMINATION

Sector I							
1a		1b		2a		2b		3		4	
$\vec{V}_{1P}$ [POO]	$\vec{V}_{2N}$ [OON]	$\vec{V}_{1P}$ [POO]	$\vec{V}_{2N}$ [OON]	$\vec{V}_{1P}$ [POO]	$\vec{V}_{2N}$ [OON]
$\vec{V}_0$ [OOO]	$\vec{V}_0$ [OOO]	$\vec{V}_7$ [PON]	$\vec{V}_7$ [PON]	$\vec{V}_7$ [PON]	$\vec{V}_7$ [PON]
$\vec{V}_{2N}$ [OON]	$\vec{V}_{1P}$ [POO]	$\vec{V}_{2N}$ [OON]	$\vec{V}_{1P}$ [POO]	$\vec{V}_{13}$ [PNN]	$\vec{V}_{14}$ [PPN]
$\vec{V}_{1N}$ [ONN]	$\vec{V}_{2P}$ [PPO]	$\vec{V}_{1N}$ [ONN]	$\vec{V}_{2P}$ [PPO]	$\vec{V}_{1N}$ [ONN]	$\vec{V}_{2P}$ [PPO]
$\vec{V}_{2N}$ [OON]	$\vec{V}_{1P}$ [POO]	$\vec{V}_{2N}$ [OON]	$\vec{V}_{1P}$ [POO]	$\vec{V}_{13}$ [PNN]	$\vec{V}_{14}$ [PPN]
$\vec{V}_0$ [OOO]	$\vec{V}_0$ [OOO]	$\vec{V}_7$ [PON]	$\vec{V}_7$ [PON]	$\vec{V}_7$ [PON]	$\vec{V}_7$ [PON]
$\vec{V}_{1P}$ [POO]	$\vec{V}_{2N}$ [OON]	$\vec{V}_{1P}$ [POO]	$\vec{V}_{2N}$ [OON]	$\vec{V}_{1P}$ [POO]	$\vec{V}_{2N}$ [OON]

Sector II							
1a		1b		2a		2b		3		4	
$\vec{V}_{2N}$ [OON]	$\vec{V}_{3P}$ [OPO]	$\vec{V}_{2N}$ [OON]	$\vec{V}_{3P}$ [OPO]	$\vec{V}_{2N}$ [OON]	$\vec{V}_{3P}$ [OPO]
$\vec{V}_0$ [OOO]	$\vec{V}_0$ [OOO]	$\vec{V}_8$ [OPN]	$\vec{V}_8$ [OPN]	$\vec{V}_8$ [OPN]	$\vec{V}_8$ [OPN]
$\vec{V}_{3P}$ [OPO]	$\vec{V}_{2N}$ [OON]	$\vec{V}_{3P}$ [OPO]	$\vec{V}_{2N}$ [OON]	$\vec{V}_{14}$ [PPN]	$\vec{V}_{15}$ [NPN]
$\vec{V}_{2P}$ [PPO]	$\vec{V}_{3N}$ [NON]	$\vec{V}_{2P}$ [PPO]	$\vec{V}_{3N}$ [NON]	$\vec{V}_{2P}$ [PPO]	$\vec{V}_{3N}$ [NON]
$\vec{V}_{3P}$ [OPO]	$\vec{V}_{2N}$ [OON]	$\vec{V}_{3P}$ [OPO]	$\vec{V}_{2N}$ [OON]	$\vec{V}_{14}$ [PPN]	$\vec{V}_{15}$ [NPN]
$\vec{V}_0$ [OOO]	$\vec{V}_0$ [OOO]	$\vec{V}_8$ [OPN]	$\vec{V}_8$ [OPN]	$\vec{V}_8$ [OPN]	$\vec{V}_8$ [OPN]
$\vec{V}_{2N}$ [OON]	$\vec{V}_{3P}$ [OPO]	$\vec{V}_{2N}$ [OON]	$\vec{V}_{3P}$ [OPO]	$\vec{V}_{2N}$ [OON]	$\vec{V}_{3P}$ [OPO]

Sector III							
1a		1b		2a		2b		3		4	
$\vec{V}_{3P}$ [OPO]	$\vec{V}_{4N}$ [NOO]	$\vec{V}_{3P}$ [OPO]	$\vec{V}_{4N}$ [NOO]	$\vec{V}_{3P}$ [OPO]	$\vec{V}_{4N}$ [NOO]
$\vec{V}_0$ [OOO]	$\vec{V}_0$ [OOO]	$\vec{V}_9$ [NPO]	$\vec{V}_9$ [NPO]	$\vec{V}_9$ [NPO]	$\vec{V}_9$ [NPO]
$\vec{V}_{4N}$ [NOO]	$\vec{V}_{3P}$ [OPO]	$\vec{V}_{4N}$ [NOO]	$\vec{V}_{3P}$ [OPO]	$\vec{V}_{15}$ [NPN]	$\vec{V}_{16}$ [NPP]
$\vec{V}_{3N}$ [NON]	$\vec{V}_{4P}$ [OPP]	$\vec{V}_{3N}$ [NON]	$\vec{V}_{4P}$ [OPP]	$\vec{V}_{3N}$ [NON]	$\vec{V}_{4P}$ [OPP]
$\vec{V}_{4N}$ [NOO]	$\vec{V}_{3P}$ [OPO]	$\vec{V}_{4N}$ [NOO]	$\vec{V}_{3P}$ [OPO]	$\vec{V}_{15}$ [NPN]	$\vec{V}_{16}$ [NPP]
$\vec{V}_0$ [OOO]	$\vec{V}_0$ [OOO]	$\vec{V}_9$ [NPO]	$\vec{V}_9$ [NPO]	$\vec{V}_9$ [NPO]	$\vec{V}_9$ [NPO]
$\vec{V}_{3P}$ [OPO]	$\vec{V}_{4N}$ [NOO]	$\vec{V}_{3P}$ [OPO]	$\vec{V}_{4N}$ [NOO]	$\vec{V}_{3P}$ [OPO]	$\vec{V}_{4N}$ [NOO]

Sector IV											
1a		1b		2a		2b		3		4	
$\vec{V}_{4N}$	[NOO]	$\vec{V}_{5P}$	[OOP]	$\vec{V}_{4N}$	[NOO]	$\vec{V}_{5P}$	[OOP]	$\vec{V}_{4N}$	[NOO]	$\vec{V}_{5P}$	[OOP]
$\vec{V}_0$	[OOO]	$\vec{V}_0$	[OOO]	$\vec{V}_{10}$	[NOP]	$\vec{V}_{10}$	[NOP]	$\vec{V}_{10}$	[NOP]	$\vec{V}_{10}$	[NOP]
$\vec{V}_{5P}$	[OOP]	$\vec{V}_{4N}$	[NOO]	$\vec{V}_{5P}$	[OOP]	$\vec{V}_{4N}$	[NOO]	$\vec{V}_{16}$	[NPP]	$\vec{V}_{17}$	[NNP]
$\vec{V}_{4P}$	[OPP]	$\vec{V}_{5N}$	[NNO]	$\vec{V}_{4P}$	[OPP]	$\vec{V}_{5N}$	[NNO]	$\vec{V}_{4P}$	[OPP]	$\vec{V}_{5N}$	[NNO]
$\vec{V}_{5P}$	[OOP]	$\vec{V}_{4N}$	[NOO]	$\vec{V}_{5P}$	[OOP]	$\vec{V}_{4N}$	[NOO]	$\vec{V}_{16}$	[NPP]	$\vec{V}_{17}$	[NNP]
$\vec{V}_0$	[OOO]	$\vec{V}_0$	[OOO]	$\vec{V}_{10}$	[NOP]	$\vec{V}_{10}$	[NOP]	$\vec{V}_{10}$	[NOP]	$\vec{V}_{10}$	[NOP]
$\vec{V}_{4N}$	[NOO]	$\vec{V}_{5P}$	[OOP]	$\vec{V}_{4N}$	[NOO]	$\vec{V}_{5P}$	[OOP]	$\vec{V}_{4N}$	[NOO]	$\vec{V}_{5P}$	[OOP]
Sector V											
1a		1b		2a		2b		3		4	
$\vec{V}_{5P}$	[OOP]	$\vec{V}_{6N}$	[ONO]	$\vec{V}_{5P}$	[OOP]	$\vec{V}_{6N}$	[ONO]	$\vec{V}_{5P}$	[OOP]	$\vec{V}_{6N}$	[ONO]
$\vec{V}_0$	[OOO]	$\vec{V}_0$	[OOO]	$\vec{V}_{11}$	[ONP]	$\vec{V}_{11}$	[ONP]	$\vec{V}_{11}$	[ONP]	$\vec{V}_{11}$	[ONP]
$\vec{V}_{6N}$	[ONO]	$\vec{V}_{5P}$	[OOP]	$\vec{V}_{6N}$	[ONO]	$\vec{V}_{5P}$	[OOP]	$\vec{V}_{17}$	[NNP]	$\vec{V}_{18}$	[PNP]
$\vec{V}_{5N}$	[NNO]	$\vec{V}_{6P}$	[POP]	$\vec{V}_{5N}$	[NNO]	$\vec{V}_{6P}$	[POP]	$\vec{V}_{5N}$	[NNO]	$\vec{V}_{6P}$	[POP]
$\vec{V}_{6N}$	[ONO]	$\vec{V}_{5P}$	[OOP]	$\vec{V}_{6N}$	[ONO]	$\vec{V}_{5P}$	[OOP]	$\vec{V}_{17}$	[NNP]	$\vec{V}_{18}$	[PNP]
$\vec{V}_0$	[OOO]	$\vec{V}_0$	[OOO]	$\vec{V}_{11}$	[ONP]	$\vec{V}_{11}$	[ONP]	$\vec{V}_{11}$	[ONP]	$\vec{V}_{11}$	[ONP]
V_{5P}	[OOP]	$\vec{V}_{6N}$	[ONO]	$\vec{V}_{5P}$	[OOP]	$\vec{V}_{6N}$	[ONO]	$\vec{V}_{5P}$	[OOP]	$\vec{V}_{6N}$	[ONO]
Sector VI											
1a		1b		2a		2b		3		4	
$\vec{V}_{6N}$	[ONO]	$\vec{V}_{1P}$	[POO]	$\vec{V}_{6N}$	[ONO]	$\vec{V}_{1P}$	[POO]	$\vec{V}_{6N}$	[ONO]	$\vec{V}_{1P}$	[POO]
$\vec{V}_0$	[OOO]	$\vec{V}_0$	[OOO]	$\vec{V}_{12}$	[PNO]	$\vec{V}_{12}$	[PNO]	$\vec{V}_{12}$	[PNO]	$\vec{V}_{12}$	[PNO]
$\vec{V}_{1P}$	[POO]	$\vec{V}_{6N}$	[ONO]	$\vec{V}_{1P}$	[POO]	$\vec{V}_{6N}$	[ONO]	$\vec{V}_{18}$	[PNP]	$\vec{V}_{13}$	[PNN]
$\vec{V}_{6P}$	[POP]	$\vec{V}_{1N}$	[ONN]	$\vec{V}_{6P}$	[POP]	$\vec{V}_{1N}$	[ONN]	$\vec{V}_{6P}$	[POP]	$\vec{V}_{1N}$	[ONN]
$\vec{V}_{1P}$	[POO]	$\vec{V}_{6N}$	[ONO]	$\vec{V}_{1P}$	[POO]	$\vec{V}_{6N}$	[ONO]	$\vec{V}_{18}$	[PNP]	$\vec{V}_{13}$	[PNN]
$\vec{V}_0$	[OOO]	$\vec{V}_0$	[OOO]	$\vec{V}_{12}$	[PNO]	$\vec{V}_{12}$	[PNO]	$\vec{V}_{12}$	[PNO]	$\vec{V}_{12}$	[PNO]
$\vec{V}_{6N}$	[ONO]	$\vec{V}_{1P}$	[POO]	$\vec{V}_{6N}$	[ONO]	$\vec{V}_{1P}$	[POO]	$\vec{V}_{6N}$	[ONO]	$\vec{V}_{1P}$	[POO]

Chapter 9

Other Multilevel Voltage Source Inverters

9.1 INTRODUCTION

The multilevel voltage source inverters have a variety of configurations. In addition to the multilevel cascaded H-bridge (CHB) inverters and neutral-point clamped (NPC) inverters discussed earlier, this chapter presents another two multilevel inverter topologies. One is the NPC/H-bridge inverter that is evolved from the three-level NPC inverter, and the other is the multilevel flying-capacitor inverter that is derived from the two-level inverter topology. The operating principle of these inverters is discussed, and their harmonic performance is analyzed.

9.2 NPC/H-BRIDGE INVERTER

The NPC/H-bridge inverter is developed from the three-level NPC inverter topology. This inverter has some unique features that have promoted its application in the MV drive industry [1].

9.2.1 Inverter Topology

The output voltage and power of a three-level NPC inverter discussed in Chapter 8 can be doubled by using 24 active switches, every two of which are connected in series. The NPC/H-bridge inverter shown in Fig. 9.2-1 also uses 24 active switches to achieve the same voltage and power ratings as the 24-switch NPC inverter. Each of the inverter phases is composed of two NPC legs in an H-bridge form.

The NPC/H-bridge inverter has some advantages over the three-level NPC inverter. The **inverter phase voltages**, v_{AN}, v_{BN} and v_{CN}, contain five voltage levels instead of three levels for the NPC inverter, leading to a lower dv/dt and THD. The inverter does not have any switching devices in series, which eliminates the device dynamic and static voltage sharing problems. However, the inverter re-

High-Power Converters and ac Drives. By Bin Wu
© 2006 The Institute of Electrical and Electronics Engineers, Inc.

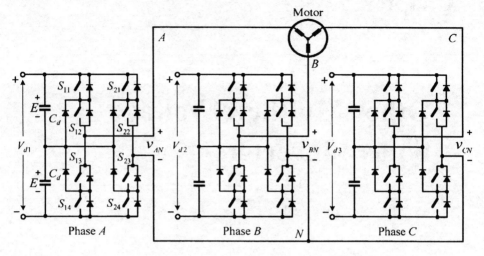

Figure 9.2-1 Five-level NPC/H-bridge inverter.

quires three isolated dc supplies, which increases the complexity and cost of the dc supply system.

9.2.2 Modulation Scheme

Figure 9.2-2 shows a modified in-phase disposition (IPD) modulation scheme for the five-level NPC/H-bridge inverter, where only the waveforms for the inverter phase A are given. The two modulating waves, v_{m1} and v_{m2}, have the same frequency and amplitude, but are 180° out of phase. Similar to the IPD modulation presented in Chapter 7, the triangular carriers v_{cr1} and v_{cr2} are in phase but vertically disposed. The frequency modulation index is defined by $m_f = f_{cr}/f_m$, where f_m and f_{cr} are the frequencies of the modulating and carrier waves. The amplitude modulation index is given by $m_a = \hat{V}_m/\hat{V}_{cr}$, where $\hat{V}_m$ and $\hat{V}_{cr}$ are the peak amplitudes of the modulating and carrier waves, respectively.

The eight switches in inverter phase A constitute four complementary switch pairs: (S_{11}, S_{13}), (S_{12}, S_{14}), (S_{21}, S_{23}), and (S_{22}, S_{24}). Therefore, the modulation scheme needs to generate only four independent gate signals for the top and bottom four switches. The modulating wave v_{m1} is used to generate the gatings v_{g11} and v_{g14} for S_{11} and S_{14}, where v_{g11} is generated during the positive half-cycle of v_{m1} while v_{g14} is produced during the negative half-cycle of v_{m1}. The gate signals for S_{21} and S_{24} are arranged in a similar manner. The waveform for the inverter phase voltage v_{AN} is formed with five voltage levels.

Unlike the IPD modulation for three-level NPC inverter where the conduction angle of the switching devices over a fundamental-frequency cycle is not the same, the gating arrangements for the five-level NPC/H-bridge inverter ensure an equal conduction angle for all the switches ($m_f > 6$). This facilitates inverter thermal de-

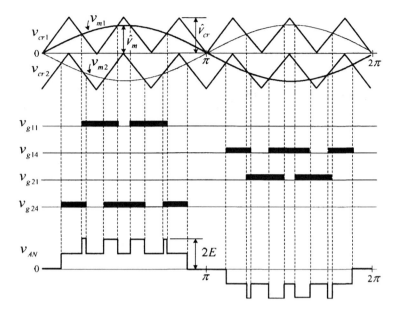

Figure 9.2-2 Modified IPD modulation for the five-level NPC/H-bridge inverter ($m_f = 6$).

sign and device selection as well. The phase-shifted multicarrier modulation discussed in Chapter 7 does not work for the NPC/H-bridge inverter.

There are three neutral points in the inverter, whose potential may need a tight control to avoid any voltage deviation. However, for the drive system using a multipulse diode rectifier as a front end, the rectifier can be designed such that its midpoints can be directly connected to the neutral points of the inverter. In doing so, the inverter neutral voltages are fixed by the rectifier and thus do not vary with the inverter operating conditions.

9.2.3 Waveforms and Harmonic Content

Figure 9.2-3 shows the simulated waveform for the phase voltage v_{AN} of the NPC/H-bridge inverter and its harmonic content. The inverter operates under the condition of $f_m = 60$ Hz, $m_f = 18$, and $m_a = 0.9$. The device switching frequency can be found from $f_{sw,dev} = f_m \times m_f/2 = 540$ Hz. The waveform of v_{AN} is composed of five voltage levels, whose harmonics appear as sidebands centered around $2m_f$ and its multiples such as $4m_f$. The phase voltage v_{AN} does not contain any harmonics with the order lower than the 27th, but has triplen harmonics such as $(2m_f \pm 3)$ and $(4m_f \pm 3)$.

The simulated waveform for the inverter line-to-line voltage v_{AB} is illustrated in Fig. 9.2-4. It contains nine voltage levels. The triplen harmonics in v_{AN} do not appear in v_{AB} due to the three-phase balanced system, resulting in a reduction of THD from 33.1% to 28.4%.

182 Chapter 9 Other Multilevel Voltage Source Inverters

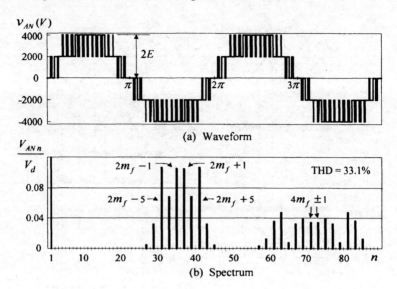

Figure 9.2-3 Waveform and spectrum of the inverter phase voltage (f_m = 60 Hz, $f_{sw,dev}$ = 540 Hz, m_f = 18, and m_a = 0.9).

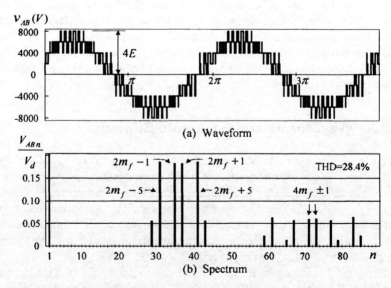

Figure 9.2-4 Waveform and spectrum of the inverter line-to-line voltage (f_m = 60 Hz, $f_{sw,dev}$ = 540 Hz, m_f = 18, and m_a = 0.9).

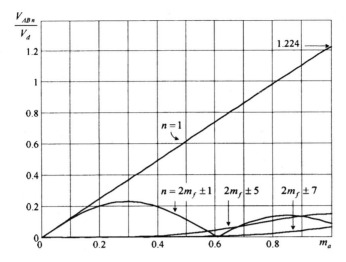

Figure 9.2-5 Harmonic content of the five-level NPC/H-bridge inverter ($m_f = 18$).

The frequency of the dominant harmonics in inverter output voltages represents the equivalent **inverter switching frequency** $f_{sw,inv}$. Since the dominant harmonics in v_{AN} and v_{AB} are distributed around $2m_f$, the inverter switching frequency can be calculated by $f_{sw,inv} = f_m \times 2m_f = 4f_{sw,dev}$, four times as high as the device switching frequency.

The harmonic content of v_{AB} versus modulation index m_a is shown in Fig. 9.2-5. Since the high-order harmonic components can be easily attenuated by filters or load inductances, only the dominant harmonics centered around $2m_f$ are plotted. The nth-order harmonic voltage V_{ANn} (rms) is normalized with respect to the dc voltage V_d. The maximum fundamental-frequency voltage at $m_a = 1$ is given by $V_{AB1,max} = 1.224 V_d$, which is two times that of the three-level NPC inverter.

9.3 MULTILEVEL FLYING-CAPACITOR INVERTERS

9.3.1 Inverter Configuration

Figure 9.3-1 shows a typical configuration of a five-level flying-capacitor inverter [2, 3]. It is evolved from the two-level inverter by adding dc capacitors to the cascaded switches. There are four complementary switch pairs in each of the inverter legs. For example, the switch pairs in leg A are (S_1, S_1'), (S_2, S_2'), (S_3, S_3'), and (S_4, S_4'). Therefore, only four independent gate signals are required for each inverter phase.

The flying-capacitor inverter in Fig. 9.3-1 can produce an inverter phase voltage with five voltage levels. When switches S_1, S_2, S_3, and S_4 conduct, the inverter

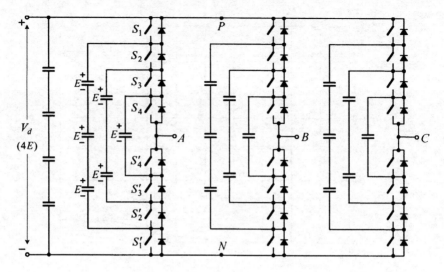

Figure 9.3-1 Five-level flying-capacitor inverter.

phase voltage v_{AN} is $4E$, which is the voltage at the inverter terminal A with respect to the negative dc bus N. Similarly, with S_1, S_2, and S_3 switched on, $v_{AN} = 3E$. Table 9.3-1 lists all the voltage levels and their corresponding switching states. It is noted that some voltage levels can be obtained by more than one switching state. The voltage level $2E$, for instance, can be produced by six sets of different switching states. The switching state redundancy is a common phenomenon in multilevel converters, which provides a great flexibility for the switching pattern design.

9.3.2 Modulation Schemes

Both phase- and level-shifted modulation schemes can be implemented for the multilevel flying-capacitor inverters. Figure 9.3-2 shows the simulated waveforms for the phase voltage v_{AN} and line-to-line voltage v_{AB} of the five-level inverter operating at the fundamental frequency of 60 Hz with a phase-shifted modulation scheme ($m_f = 12$, $m_a = 0.8$, $f_{sw,dev} = 720$ Hz, and $f_{sw,inv} = 2880$ Hz). The equivalent inverter switching frequency $f_{sw,inv}$ is four times the device switching frequency $f_{sw,dev}$.

Since the flying-capacitor inverter topology is derived from the two-level inverter, it carries the same features as the two-level inverter such as modular structure for the switching devices. It is also a multilevel inverter, producing the voltage waveforms with reduced dv/dt and THD. However, the flying-capacitor inverter has some limitations, including the following:

- A large number of dc capacitors with separate pre-charge circuits. The inverter requires several banks of bulky dc capacitors, each of which needs a separate pre-charge circuit.

9.3 Multilevel Flying-Capacitor Inverters

Table 9.3-1 Voltage Level and Switching State of a Five-Level Flying-Capacitor Inverter

Inverter Phase Voltage V_{AN}	Switching State			
	S_1	S_2	S_3	S_4
$4E$	1	1	1	1
$3E$	1	1	1	0
	0	1	1	1
	1	0	1	1
	1	1	0	1
$2E$	1	1	0	0
	0	0	1	1
	1	0	0	1
	0	1	1	0
	1	0	1	0
	0	1	0	1
$1E$	1	0	0	0
	0	1	0	0
	0	0	1	0
	0	0	0	1
0	0	0	0	0

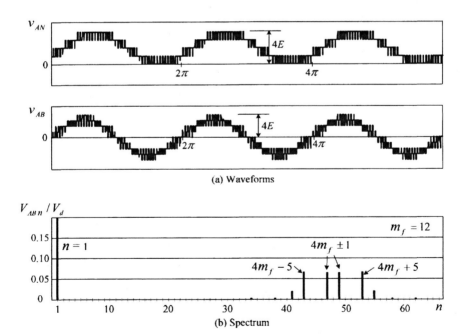

Figure 9.3-2 Waveforms and spectrum of the five-level flying-capacitor inverter.

- Complex capacitor voltage balancing control. The dc capacitor voltages in the inverter normally vary with the inverter operating conditions. To avoid the problems caused by the dc voltage deviation, the voltages on the dc flying capacitors should be tightly controlled, which increases the complexity of the control scheme.

Due to the above-mentioned drawbacks, the practical use of the flying-capacitor inverter in the drive system seems limited.

9.4 SUMMARY

The chapter focuses on two multilevel inverter topologies that are not covered in the previous chapters. One is the five-level NPC/H-bridge inverter and the other is the multilevel flying-capacitor inverter. The operating principle of these inverters is explained, and their modulation schemes are discussed. The NPC/H-bridge inverter has some unique features that have promoted its application in the MV drives, whereas the practical use of the flying-capacitor inverter seems limited due to the use of a large number of capacitors and complex control scheme.

REFERENCES

1. GE Toshiba Automation Systems, *A New Family of MV Drives for a New Century—DURA BILT 5i MV*, Product Brochure, 50 pages, March 2003.
2. M. F. Escalante, J. C. Vannier, et al., Flying Capacitor Multilevel Inverters and DTC Motor Drive Applications, *IEEE Transactions on Industrial Applications*, Vol. 49, No. 4, pp. 809–815, 2002.
3. L. Xu and V. G. Agelidis, Flying Capacitor Multilevel PWM Converter Based UPFC, *IEE Proceedings on Electric Power Applications*, Vol. 149, No. 4, pp. 304–310, 2002.

Part Four

PWM Current Source Converters

Chapter 10

PWM Current Source Inverters

10.1 INTRODUCTION

The inverters used in medium-voltage (MV) drives can be generally classified into voltage source inverter (VSI) and current source inverter (CSI). The voltage source inverter produces a defined three-phase PWM voltage waveform for the load while the current source inverter outputs a defined PWM current waveform. The PWM current source inverter features simple converter topology, motor-friendly waveforms, and reliable short-circuit protection, and therefore it is one of the widely used converter topologies for the MV drive [1].

Two types of current source inverters are commonly used in the MV drive: PWM inverters and load-commutated inverter (LCI). The PWM inverter uses switching devices with self-extinguishable capability. Prior to the advent of GCT devices in the late 1990s, GTOs were dominantly used in the CSI-fed drives [2, 3]. The load-commutated inverter employs SCR thyristors whose commutation is assisted by the load with a leading power factor. The LCI topology is particularly suitable for very large synchronous motor drives with a power rating up to 100 MW [4].

This chapter mainly deals with the PWM current source inverter. Various modulation techniques for the inverter are discussed, which include trapezoidal pulse width modulation (TPWM), selective harmonic elimination (SHE), and space vector modulation (SVM). These modulation schemes are developed for high-power inverters operating with a switching frequency of around 500 Hz. The operating principle of the modulation schemes is elaborated and their harmonic performance is analyzed. A new CSI topology using parallel inverters is also presented. An SVM-based dc current balance control algorithm is proposed for the parallel inverters. The chapter ends with an introduction to the load-commutated inverter.

10.2 PWM CURRENT SOURCE INVERTER

An idealized PWM current source inverter is shown in Fig. 10.2-1. The inverter is composed of six GCT devices, each of which can be replaced with two or more devices in series for medium-voltage operation. The GCT devices used in the current source inverter are of symmetrical type with a reverse voltage blocking capability. The inverter produces a defined PWM output current i_w. On the dc side of the inverter is an ideal dc current source I_d. In practice, I_d can be obtained by a current source rectifier (CSR).

The current source inverter normally requires a three-phase capacitor C_f at its output to assist the commutation of the switching devices. For instance, at the turn-off of switch S_1, the inverter PWM current i_w falls to zero within a very short period of time. The capacitor provides a current path for the energy trapped in the phase-A load inductance. Otherwise, a high-voltage spike would be induced, causing damages to the switching devices. The capacitor also acts as a harmonic filter, improving the load current and voltage waveforms. The value of capacitor is in the range of 0.3 to 0.6 per unit for the MV drive with a switching frequency of around 200 Hz [5]. The capacitor value can be reduced accordingly with the increase of the switching frequency.

The dc current source I_d can be realized by an SCR or PWM current source rectifier with a dc current feedback control as shown in Fig. 10.2-2. To make the dc current I_d smooth and continuous, a dc choke L_d is an indispensable device for the current source rectifier. Through the feedback control the magnitude of I_d is kept at a value set by its reference I_d^*. The size of the dc choke is normally in the range of 0.5 to 0.8 per unit.

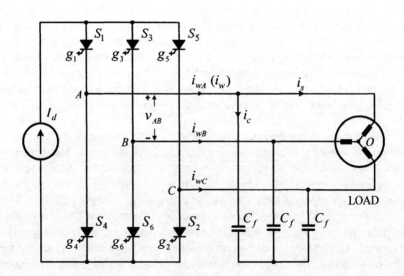

Figure 10.2-1 PWM GCT current source inverter.

10.2 PWM Current Source Inverter

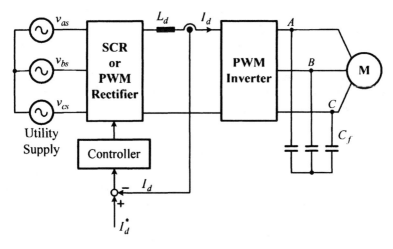

Figure 10.2-2 Realization of a dc current source I_d.

The PWM current source inverter has the following characteristics.

- **Simple Converter Topology.** The GCT devices used in the inverter are of symmetrical type, which do not require antiparallel freewheeling diodes.
- **Motor Friendly Waveforms.** The current source inverter produces a three-phase PWM current instead of PWM voltage as in the VSI. With the filter capacitor installed at the inverter output, the load current and voltage waveforms are close to sinusoidal. The high dv/dt problem associated with the VSI does not exist in the CSI.
- **Reliable Short-Circuit Protection.** In case of a short circuit at the inverter output terminals, the rate of rise of the dc current is limited by the dc choke, allowing sufficient time for the protection circuit to function.
- **Limited Dynamic Performance.** The dc current cannot be changed instantaneously during transients, which reduces the system dynamic performance.

10.2.1 Trapezoidal Modulation

The switching pattern design for the CSI should generally satisfy two conditions: (a) The dc current I_d should be continuous, and (b) the inverter PWM current i_w should be defined. The two conditions can be translated into a **switching constraint**: At any instant of time (excluding commutation intervals) there are only two switches conducting, one in the top half of the bridge and the other in the bottom half. With only one switch turned on, the continuity of the dc current is lost. A very high voltage will be induced by the dc choke, causing damage to the switching devices. If more than two devices are on simultaneously, the PWM current i_w is not defined by the switching pattern. For instance, with S_1, S_2 and S_3 conducting at the same time, the currents in S_1 and S_3, which are the PWM currents in the inverter phases A and B, are load-dependent although the sum of the two currents is equal to I_d.

Figure 10.2-3 shows the principle of trapezoidal pulse width modulation (TPWM), where v_m is a trapezoidal modulating wave and v_{cr} is a triangular carrier wave. The **amplitude modulation index** is defined by

$$m_a = \frac{\hat{V}_m}{\hat{V}_{cr}} \qquad (10.2\text{-}1)$$

where $\hat{V}_m$ and $\hat{V}_{cr}$ are the peak values of the modulating and carrier waves, respectively. Similar to the carrier based PWM schemes for voltage source inverters, the gate signal v_{g1} for switch S_1 is generated by comparing v_m with v_{cr}. However, the trapezoidal modulation does not generate gatings in the center $\pi/3$ interval of the positive half-cycle or in the negative half-cycle of the inverter fundamental frequency. Such an arrangement leads to the satisfaction of the switching constraint for the CSI. It can be observed from the gate signals that only two GCTs conduct at any time, resulting in a defined i_w. The magnitude of i_w is set by the dc current I_d.

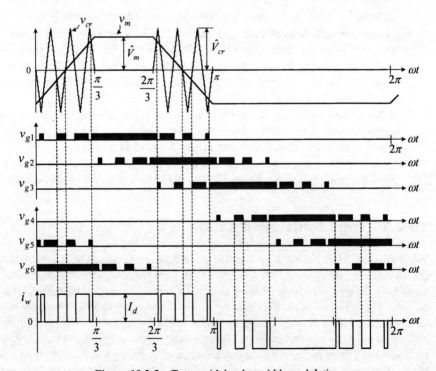

Figure 10.2-3 Trapezoidal pulse-width modulation.

10.2 PWM Current Source Inverter

The switching frequency of the devices can be calculated by

$$f_{sw} = f_1 \times N_p \tag{10.2-2}$$

where f_1 is the fundamental frequency and N_p is the number of pulses per half-cycle of i_w.

Figure 10.2-4a shows the spectrum of the inverter PWM current i_w with $N_p = 13$ and $m_a = 0.83$, where I_{wn} is the rms value of the nth-order harmonic current in i_w and $I_{w1,\max}$ is the maximum rms fundamental-frequency current given by

$$I_{w1,\max} = 0.74 I_d \quad \text{for } m_a = 1 \tag{10.2-3}$$

The PWM current i_w does not contain any even-order harmonics since its waveform is of half-wave symmetry. The TPWM scheme produces two pairs of dominant harmonics at $n = 3(N_p - 1) \pm 1$ and $n = 3(N_p - 1) \pm 5$, which are the 31st, 35th, 37th, and 41st in this case.

The harmonic content in i_w is shown in Fig. 10.2-4b. Its fundamental-frequency component I_{w1} does not vary significantly with the modulation index m_a. When m_a varies from zero to its maximum value of 1.0, I_{w1} changes from its minimum value of $0.89 I_{w1,\max}$ to $I_{w1,\max}$, presenting only an 11% increase. This is mainly due to the fact that i_w is not modulated in the center $\pi/3$ interval of each half-cycle. In practice, the adjustment of I_{w1} is normally accomplished by varying the dc current I_d through the rectifier instead of m_a.

Figure 10.2-4c shows the details of the low-order harmonic currents. At the modulation index of 0.85, the magnitudes of most harmonic currents are close to their lowest values, leading to a lowest harmonic distortion. This phenomenon holds true for other values of N_p. Therefore, the modulation index of 0.85 can be selected, at which the THD of i_w is minimized while the fundamental current I_{w1} is close to $I_{w1,\max}$.

Figure 10.2-5 shows the experimental results obtained from a low-power laboratory CSI-fed induction motor drive with $m_a = 0.85$ and $C_f = 0.66$ pu. The motor operated at low speeds with a rated stator current. The oscillogram in Fig. 10.2-5a illustrates the waveforms for the inverter PWM current i_w, stator current i_s and line-to-line voltage v_{AB} with the fundamental frequency $f_1 = 13.8$ Hz and switching frequency $f_{sw} = 180$ Hz. The PWM current i_w contains 13 pulses per half-cycle ($N_p = 13$). The waveform for i_s looks like a trapezoid superimposed with some switching noise. Although the v_{AB} waveform contains more harmonics than i_s, it is much better than that of the two-level VSI in terms of harmonic distortion and dv/dt. The measured waveforms for the inverter operating at $f_1 = 5$ Hz and $f_{sw} = 155$ Hz ($N_p = 31$) are shown in Fig. 10.2-5b. Considering the fact that the switching frequency in these two cases is only 180 Hz and 155 Hz, the waveforms for v_{AB} and i_s are fairly good.

As indicated earlier, the TPWM scheme produces two pairs of dominant harmonics at $n = 3(N_p - 1) \pm 1$ and $n = 3(N_p - 1) \pm 5$. With $N_p = 5$, the dominant harmonics in i_w include the 7th, 11th, 13th, and 17th. These low-order harmonics are

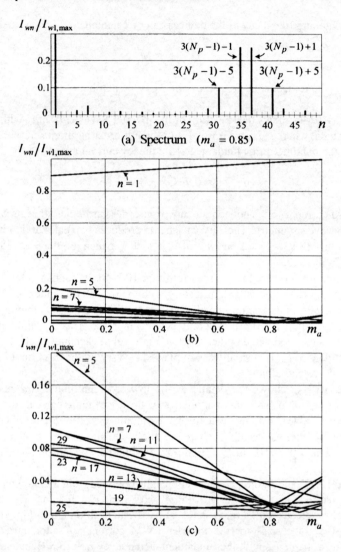

Figure 10.2-4 Harmonic content of i_w produced by TPWM with $N_p = 13$.

difficult to be fully attenuated by the filter capacitor and motor inductance, causing a detrimental effect on motor operation and harmonic power losses. Therefore, the applicability of the trapezoidal modulation is diminished for $N_p \leq 7$.

10.2.2 Selective Harmonic Elimination

Selective harmonic elimination (SHE) is an off-line modulation scheme, which is able to eliminate a number of low-order unwanted harmonics in the inverter PWM

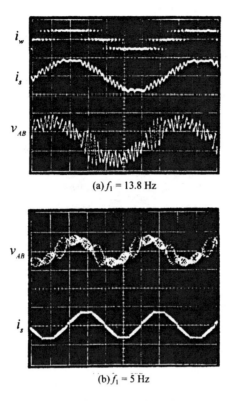

(a) $f_1 = 13.8$ Hz

(b) $f_1 = 5$ Hz

Figure 10.2-5 Waveforms measured from a CSI drive with TPWM scheme. (a) $f_1 = 13.8$ Hz, $N_p = 13$, $f_{sw} = 180$ Hz. (b) $f_1 = 5$ Hz, $N_p = 31$, $f_{sw} = 155$ Hz.

current i_w. The switching angles are pre-calculated and then imported into a digital controller for implementation. Figure 10.2-6 shows a typical SHE waveform that satisfies the switching constraint for the CSI. There are five pulses per half-cycle ($N_p = 5$) with five switching angles in the first $\pi/2$ period. However, only two out of the five angles, θ_1 and θ_2, are independent. Given these two angles, all other switching angles can be calculated.

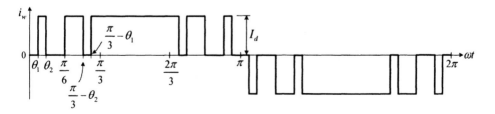

Figure 10.2-6 Selective harmonic elimination (SHE) scheme.

The two switching angles provide two degrees of freedom, which can be used to either eliminate two harmonics in i_w without modulation index control or eliminate one harmonic and provide an adjustable modulation index m_a. The first option is preferred since the adjustment of I_{w1} is normally done by varying I_d through the rectifier. The number of harmonics to be eliminated is then given by

$$k = (N_p - 1)/2$$

The inverter PWM current i_w can generally be expressed as

$$i_w(\omega t) = \sum_{n=1}^{\infty} a_n \sin(n\omega t) \tag{10.2-4}$$

where

$$a_n = \frac{4}{\pi} \int_0^{\frac{\pi}{2}} i_w(\omega t)\sin(n\omega t)d(\omega t) \tag{10.2-5}$$

The Fourier coefficient a_n can be found from

$$a_n = \frac{4I_{dc}}{\pi} \times \begin{cases} \int_{\theta_1}^{\theta_2} \sin(n\omega t)d(\omega t) + \cdots + \int_{\theta_k}^{\frac{\pi}{6}} \sin(n\omega t)d(\omega t) \\ + \int_{\frac{\pi}{3}-\theta_k}^{\frac{\pi}{3}-\theta_{k-1}} \sin(n\omega t)d(\omega t) + \cdots + \int_{\frac{\pi}{3}-\theta_1}^{\frac{\pi}{2}} \sin(n\omega t)d(\omega t), & k = \text{odd}; \\ \int_{\theta_1}^{\theta_2} \sin(n\omega t)d(\omega t) + \cdots + \int_{\theta_{k-1}}^{\theta_k} \sin(n\omega t)d(\omega t) \\ + \int_{\frac{\pi}{6}}^{\frac{\pi}{3}-\theta_k} \sin(n\omega t)d(\omega t) + \cdots + \int_{\frac{\pi}{3}-\theta_1}^{\frac{\pi}{2}} \sin(n\omega t)d(\omega t), & k = \text{even} \end{cases} \tag{10.2-6}$$

from which

$$a_n = \frac{4I_{dc}}{\pi n} \times \begin{cases} \cos(n\theta_1) + \cos[n(\pi/3 - \theta_1)] - \cos(n\theta_2) - \cos[n(\pi/3 - \theta_2)] + \cdots \\ + \cos(n\theta_k) + \cos[n(\pi/3 - \theta_k)] - \cos(n\pi/6), & k = \text{odd}; \\ \cos(n\theta_1) + \cos[n(\pi/3 - \theta_1)] - \cos(n\theta_2) - \cos[n(\pi/3 - \theta_2)] + \cdots \\ - \cos(n\theta_k) - \cos[n(\pi/3 - \theta_k)] + \cos(n\pi/6), & k = \text{even} \end{cases} \tag{10.2-7}$$

To eliminate k harmonics, k equations can be formulated by setting $a_n = 0$,

$$F_i = (\theta_1, \theta_2, \theta_3, \ldots, \theta_k) = 0, \quad i = 1, 2, \ldots, k \tag{10.2-8}$$

For example, to eliminate the 5th, 7th and 11th harmonics in i_w, the following three functions can be derived

$$F_1 = \cos(5\theta_1) + \cos[5(\pi/3 - \theta_1)] - \cos(5\theta_2) - \cos[5(\pi/3 - \theta_2)]$$
$$+ \cos(5\theta_3) + \cos[5(\pi/3 - \theta_3)] - \cos(5\pi/6) = 0$$

$$F_2 = \cos(7\theta_1) + \cos[7(\pi/3 - \theta_1)] - \cos(7\theta_2) - \cos[7(\pi/3 - \theta_2)]$$
$$+ \cos(7\theta_3) + \cos[7(\pi/3 - \theta_3)] - \cos(7\pi/6) = 0 \qquad (10.2\text{-}9)$$

$$F_3 = \cos(11\theta_1) + \cos[11(\pi/3 - \theta_1)] - \cos(11\theta_2) - \cos[11(\pi/3 - \theta_2)]$$
$$+ \cos(11\theta_3) + \cos[11(\pi/3 - \theta_3)] - \cos(11\pi/6) = 0$$

The nonlinear and transcendental equation of (10.2-9) can be solved by a number of numerical methods, one of which is the Newton–Raphson iteration algorithm [6]. The flowchart of this algorithm is shown in Fig. 10.2-7, where θ^0 is the initial guess of the switching angles and $\partial F/\partial \theta$ is the Jacob matrix given by

$$\frac{\partial F}{\partial \theta} = \begin{bmatrix} \dfrac{\partial F_1}{\partial \theta_1} & \dfrac{\partial F_1}{\partial \theta_2} & \dfrac{\partial F_1}{\partial \theta_3} & \cdots & \dfrac{\partial F_1}{\partial \theta_k} \\ \dfrac{\partial F_2}{\partial \theta_1} & \dfrac{\partial F_2}{\partial \theta_2} & \dfrac{\partial F_2}{\partial \theta_3} & \cdots & \dfrac{\partial F_2}{\partial \theta_k} \\ \cdots & \cdots & \cdots & \cdots & \cdots \\ \dfrac{\partial F_k}{\partial \theta_1} & \dfrac{\partial F_k}{\partial \theta_2} & \dfrac{\partial F_k}{\partial \theta_3} & \cdots & \dfrac{\partial F_k}{\partial \theta_k} \end{bmatrix} \qquad (10.2\text{-}10)$$

Based on the algorithm, the switching angles for the 5th, 7th, and 11th harmonic elimination are $\theta_1 = 2.24°$, $\theta_2 = 5.60°$, and $\theta_3 = 21.26°$. The PWM current i_w and its spectrum are illustrated in Fig. 10.2-8. Similar to the TPWM scheme, the SHE modulation generates two pairs of dominant harmonics at $n = 3(N_p - 1) \pm 1$ and $n = 3(N_p - 1) \pm 5$.

The switching angles for the elimination of up to four harmonics in i_w are given in the appendix of this chapter. It is worth noting that the harmonics with the lowest order normally have the highest priority to be eliminated, but it is not always the case. For example, if an LC resonant mode, caused by the filter capacitor and the load inductance, happens to be the frequency of the 11th harmonics, the 5th and 11th harmonics should be eliminated to avoid any possible resonances instead of the 5th and 7th harmonic elimination. In addition, Eq. (10.2-8) may not always have a valid solution. For example, no valid solutions can be found to eliminate the 5th, 7th, 11th, 13th, and 17th simultaneously.

Figure 10.2-9 shows a set of voltage and current waveforms measured from a low-horsepower CSI-fed induction motor drive operating at various speeds with C_f = 0.66 pu [5]. To keep the motor air-gap flux constant, the magnitude of the stator

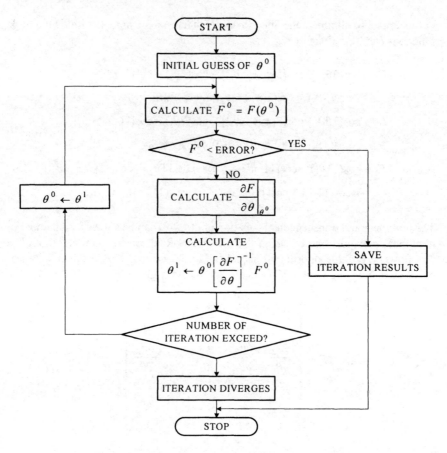

Figure 10.2-7 Flowchart of Newton–Raphson algorithm.

voltage v_{AB} changes with the inverter fundamental frequency f_1 accordingly while the stator current i_s is kept at its rated value.

The oscillogram in Fig. 10.2-9a shows the waveforms of the inverter operating at 20 Hz, where three low-order harmonics, the 5th, 7th, and 11th, are eliminated. The resultant switching frequency is only 140 Hz. When the inverter operates at 35 Hz as illustrated in Fig. 10.2-9b with the 5th and 7th harmonics eliminated, its switching frequency is 175 Hz. The waveforms of the inverter working at 60 Hz is shown in Fig. 10.2-9c, where only the 5th harmonic current is eliminated, leading to a switching frequency of 180 Hz.

It can be observed that the current source inverter produces motor-friendly waveforms even with a low switching frequency ($\leq$ 180 Hz). The high dv/dt problems associated with the voltage source inverters normally do not exist in the current source inverters.

The SHE and TPWM modulation techniques presented above can be combined for high-power MV drives, where the SHE modulation can be used when the invert-

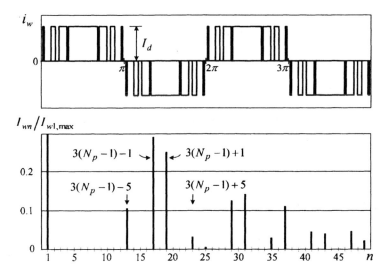

Figure 10.2-8 Waveforms of i_w and its spectrum with 5th, 7th, and 11th harmonic elimination.

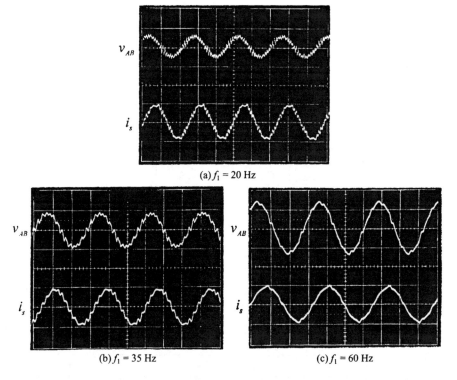

Figure 10.2-9 Waveforms measured from a CSI-fed drive using SHE scheme. (a) $f_1 = 20$ Hz, $f_{sw} = 140$ Hz. (b) $f_1 = 35$ Hz, $f_{sw} = 175$ Hz. (c) $f_1 = 60$ Hz, $f_{sw} = 180$ Hz.

er operates at high fundamental frequencies whereas the TPWM scheme can be utilized for the inverter running at low frequencies. This topic will be further discussed in Chapter 13.

10.3 SPACE VECTOR MODULATION

In addition to the TPWM and SHE schemes, the current source inverter can also be controlled by space vector modulation (SVM) [7, 8]. In this section, the principle and implementation of the SVM scheme are presented and its performance is compared with the TPWM and SHE modulation techniques.

10.3.1 Switching States

As stated earlier, the PWM switching pattern for the CSI shown in Fig. 10.2-1 must satisfy a constraint, that is, only two switches in the inverter conduct at any time instant, one in the top half of the CSI bridge and the other in the bottom half. Under this constraint, the three-phase inverter has a total of nine switching states as listed in Table 10.3-1. These switching states can be classified as **zero switching states** and **active switching states**.

There are three zero switching states [14], [36], and [52]. The zero state [14] signifies that switches S_1 and S_4 in inverter phase leg A conduct simultaneously and the other four switches in the inverter are off. The dc current source I_d is bypassed, leading to $i_{wA} = i_{wB} = i_{wC} = 0$. This operating mode is often referred to as **bypass operation**.

There exist six active switching states. State [12] indicates that switch S_1 in leg A and S_2 in leg C are on. The dc current I_d flows through S_1, the load, S_2, and then

Table 10.3-1 Switching States and Space Vectors

Type	Switching State	On-State Switch	Inverter PWM Current			Space Vector
			i_{wA}	i_{wB}	i_{wC}	
Zero States	[14]	S_1, S_4	0	0	0	$\vec{I}_0$
	[36]	S_3, S_6				
	[52]	S_5, S_2				
Active States	[61]	S_6, S_1	I_d	$-I_d$	0	$\vec{I}_1$
	[12]	S_1, S_2	I_d	0	$-I_d$	$\vec{I}_2$
	[23]	S_2, S_3	0	I_d	$-I_d$	$\vec{I}_3$
	[34]	S_3, S_4	$-I_d$	I_d	0	$\vec{I}_4$
	[45]	S_4, S_5	$-I_d$	0	I_d	$\vec{I}_5$
	[56]	S_5, S_6	0	$-I_d$	I_d	$\vec{I}_6$

back to the dc source, resulting in $i_{wA} = I_d$ and $i_{wC} = -I_d$. The definition of other five active states is also given in the table.

10.3.2 Space Vectors

The active and zero switching states can be represented by active and zero space vectors, respectively. A typical space vector diagram for the CSI is shown in Fig. 10.3-1, where $\vec{I}_1$ to $\vec{I}_6$ are the **active vectors** and $\vec{I}_0$ is the **zero vector**. The active vectors form a regular hexagon with six equal sectors while the zero vector $\vec{I}_0$ lies on the center of the hexagon.

To derive the relationship between the space vectors and switching states, we can follow the same procedures given in Chapter 6. Assuming that the operation of the inverter in Fig. 10.2-1 is three-phase balanced, we have

$$i_{wA}(t) + i_{wB}(t) + i_{wC}(t) = 0 \tag{10.3-1}$$

where i_{wA}, i_{wB}, and i_{wC} are the instantaneous PWM output currents in the inverter phases A, B, and C, respectively. The three-phase currents can be transformed into two-phase currents in the α–β plane

$$\begin{bmatrix} i_\alpha(t) \\ i_\beta(t) \end{bmatrix} = \frac{2}{3} \begin{bmatrix} 1 & -\frac{1}{2} & -\frac{1}{2} \\ 0 & \frac{\sqrt{3}}{2} & -\frac{\sqrt{3}}{2} \end{bmatrix} \begin{bmatrix} i_{wA}(t) \\ i_{wB}(t) \\ i_{wC}(t) \end{bmatrix} \tag{10.3-2}$$

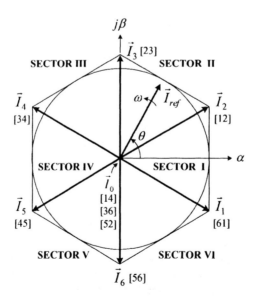

Figure 10.3-1 Space vector diagram for the current source inverter.

A current space vector can be generally expressed in terms of the two-phase currents as

$$\vec{I}(t) = i_\alpha(t) + ji_\beta(t) \tag{10.3-3}$$

Substituting (10.3-2) into (10.3-3), $\vec{I}(t)$ can be expressed in terms of i_{wA}, i_{wB}, and i_{wC}:

$$\vec{I}(t) = \frac{2}{3}[i_{wA}(t)e^{j0} + i_{wB}(t)e^{j2\pi/3} + i_{wC}(t)e^{j4\pi/3}] \tag{10.3-4}$$

For the active state [61], S_1 and S_6 are turned on and the inverter PWM currents are

$$i_{wA}(t) = I_d, \quad i_{wB}(t) = -I_d, \quad \text{and} \quad i_{wC}(t) = 0 \tag{10.3-5}$$

Substituting (10.3-5) into (10.3-4) yields

$$\vec{I}_1 = \frac{2}{\sqrt{3}} I_d e^{j(-\pi/6)} \tag{10.3-6}$$

Similarly, the other five active vectors can be derived. The active vectors can be expressed as

$$\vec{I}_k = \frac{2}{\sqrt{3}} I_d e^{j\left((k-1)\frac{\pi}{3} - \frac{\pi}{6}\right)} \quad \text{for } k = 1, 2, \ldots, 6 \tag{10.3-7}$$

Note that the active and zero vectors do not move in space, and thus they are referred to as **stationary vectors**. On the contrary, the **current reference vector** $\vec{I}_{ref}$ in Fig. 10.3-1 rotates in space at an angular velocity

$$\omega = 2\pi f_1 \tag{10.3-8}$$

where f_1 is the fundamental frequency of the inverter output current i_w. The angular displacement between $\vec{I}_{ref}$ and the α-axis of the α–β plane can be obtained by

$$\theta(t) = \int_0^t \omega(t)dt + \theta(0) \tag{10.3-9}$$

For a given length and position, $\vec{I}_{ref}$ can be synthesized by three nearby stationary vectors, based on which the switching states of the inverter can be selected and gate signals for the active switches can be generated. When $\vec{I}_{ref}$ passes through sectors one by one, different sets of switches are turned on or off. As a result, when $\vec{I}_{ref}$ rotates one revolution in space, the inverter output current varies one cycle over time.

The inverter output frequency corresponds to the rotating speed of $\vec{I}_{ref}$ while its output current can be adjusted by the length of $\vec{I}_{ref}$.

10.3.3 Dwell Time Calculation

As indicated above, the reference $\vec{I}_{ref}$ can be synthesized by three stationary vectors. The dwell time for the stationary vectors essentially represents the duty-cycle time (on-state or off-state time) of the chosen switches during a **sampling period** T_s. The dwell time calculation is based on ampere-second balancing principle, that is, the product of the reference vector $\vec{I}_{ref}$ and sampling period T_s equals the sum of the current vectors multiplied by the time interval of chosen space vectors. Assuming that the sampling period T_s is sufficiently small, the reference vector $\vec{I}_{ref}$ can be considered constant during T_s. Under this assumption, $\vec{I}_{ref}$ can be approximated by two adjacent active vectors and a zero vector. For example, with $\vec{I}_{ref}$ falling into sector I as shown in Fig. 10.3-2, it can be synthesized by $\vec{I}_1$, $\vec{I}_2$, and $\vec{I}_0$. The ampere-second balancing equation is thus given by

$$\vec{I}_{ref} T_s = \vec{I}_1 T_1 + \vec{I}_2 T_2 + \vec{I}_0 T_0$$
$$T_s = T_1 + T_2 + T_0$$
(10.3-10)

where T_1, T_2, and T_0 are the dwell times for the vectors $\vec{I}_1$, $\vec{I}_2$, and $\vec{I}_0$, respectively.
Substituting

$$\vec{I}_{ref} = I_{ref} e^{j\theta}, \quad \vec{I}_1 = \frac{2}{\sqrt{3}} I_d e^{-j\frac{\pi}{6}}, \quad \vec{I}_2 = \frac{2}{\sqrt{3}} I_d e^{j\frac{\pi}{6}}, \quad \text{and} \quad \vec{I}_0 = 0 \quad (10.3\text{-}11)$$

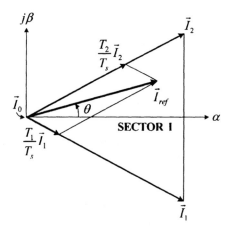

Figure 10.3-2 Synthesis of $\vec{I}_{ref}$ by $\vec{I}_1$, $\vec{I}_2$, and $\vec{I}_0$.

into (10.3-10) and then splitting the resultant equation into the real (α-axis) and imaginary (β-axis) components leads to

$$\text{Re:} \quad I_{ref}(\cos \theta)T_s = I_d(T_1 + T_2)$$

$$\text{Im:} \quad I_{ref}(\sin \theta)T_s = \frac{1}{\sqrt{3}} I_d(-T_1 + T_2) \quad (10.3\text{-}12)$$

Solving (10.3-12) together with $T_s = T_1 + T_2 + T_0$ gives

$$\begin{aligned} T_1 &= m_a \sin(\pi/6 - \theta)T_s \\ T_2 &= m_a \sin(\pi/6 + \theta)T_s \quad \text{for } -\pi/6 \leq \theta < \pi/6 \\ T_0 &= T_s - T_1 - T_2 \end{aligned} \quad (10.3\text{-}13)$$

where m_a is the modulation index, given by

$$m_a = \frac{I_{ref}}{I_d} = \frac{\hat{I}_{w1}}{I_d} \quad (10.3\text{-}14)$$

in which $\hat{I}_{w1}$ is the peak value of the fundamental-frequency component in i_w.

Note that although Eq. (10.3-13) is derived when $\vec{I}_{ref}$ is in sector I, it can also be used when $\vec{I}_{ref}$ is in other sectors provided that a multiple of $\pi/3$ is subtracted from the actual angular displacement θ such that the modified angle θ' falls into the range of $-\pi/6 \leq \theta' < \pi/6$ for use in the equation, that is,

$$\theta' = \theta - (k-1)\pi/3 \quad \text{for } -\pi/6 \leq \theta' < \pi/6 \quad (10.3\text{-}15)$$

where $k = 1, 2, \ldots, 6$ for sectors I, II, \ldots, VI, respectively.

The maximum length of the reference vector, $I_{ref,max}$, corresponds to the radius of the largest circle that can be inscribed within the hexagon as shown in Fig. 10.3-1. Since the hexagon is formed by the six active vectors having a length of $2I_d/\sqrt{3}$, $I_{ref,max}$ can be found from

$$I_{ref,max} = \frac{2I_d}{\sqrt{3}} \times \frac{\sqrt{3}}{2} = I_d \quad (10.3\text{-}16)$$

Substituting (10.3-16) into (10.3-14) gives the maximum modulation index

$$m_{a,max} = 1 \quad (10.3\text{-}17)$$

from which the modulation index is in the range of

$$0 \leq m_a \leq 1 \quad (10.3\text{-}18)$$

10.3.4 Switching Sequence

Similar to the space vector modulation for the two-level VSI, the switching sequence design for the CSI should satisfy the following two requirements for the minimization of switching frequencies:

(a) The transition from one switching state to the next involves only two switches, one being switched on and the other switched off.
(b) The transition for $\vec{I}_{ref}$ moving from one sector to the next requires the minimum number of switchings.

Figure 10.3-3 shows a typical three-segment sequence for the reference vector $\vec{I}_{ref}$ residing in sector I, where v_{g1} to v_{g6} are the gate signals for switches S_1 to S_6, respectively. The reference vector $\vec{I}_{ref}$ is synthesized by $\vec{I}_1$, $\vec{I}_2$, and $\vec{I}_0$. The sampling period T_s is divided into three segments composed of T_1, T_2, and T_0. The switching states for vectors $\vec{I}_1$ and $\vec{I}_2$ are [61] and [12], and their corresponding on-state switch pairs are (S_6, S_1) and (S_1, S_2). The zero state [14] is selected for $\vec{I}_0$ such that the design requirement (a) is satisfied.

Figure 10.3-4 shows the details of the switching sequence and gate signal arrangements over a fundamental-frequency cycle. There are 12 samples per cycle with two samples in each sector. We can observe the following:

- At any time instant, only two switches conduct, one in the top half of the bridge and the other in the bottom half.
- By a proper selection of the redundant switching states for $\vec{I}_0$, the requirements for switching sequence design are satisfied. In particular, the transition for $\vec{I}_{ref}$ moving from one sector to the next involves only two switches.

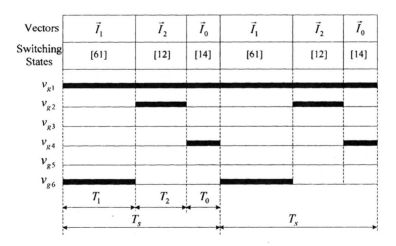

Figure 10.3-3 Switching sequence for $\vec{I}_{ref}$ in sector I.

- The dc current I_d is bypassed 12 times per fundamental-frequency cycle by the zero vector. It is the bypass operation that makes the magnitude of the fundamental-frequency current i_{w1} adjustable.
- The inverter PWM current i_w varies one cycle when the reference vector $\vec{I}_{ref}$ passes through all six sectors once.
- The device switching frequency f_{sw} can be calculated by $f_{sw} = f_1 \times N_p$.
- The sampling frequency is $f_{sp} = 1/T_s$, which relates the switching frequency by $f_{sw} = f_{sp}/2$.
- The switching sequence for the SVM scheme is

$$\begin{aligned} \vec{I}_k, \vec{I}_{k+1}, \vec{I}_0 & \quad \text{for } k = 1, 2, \ldots, 5 \\ \vec{I}_k, \vec{I}_1, \vec{I}_0 & \quad \text{for } k = 6 \end{aligned} \quad (10.3\text{-}19)$$

where k represents the sector number.

The simulated waveforms for a 1MV/4160V CSI using the space vector modulation is shown in Fig. 10.3-5, where i_w is the inverter PWM current, i_s is the load phase current, and v_{AB} is the inverter line-to-line voltage. The inverter operates at f_1 = 60 Hz, f_{sp} = 1080 Hz, and f_{sw} = 540 Hz with $m_a = 1$. The filter capacitor C_f is 0.3 pu per phase. The inverter is loaded with a three-phase balanced inductive load having a resistance of 1.0 pu and an inductance of 0.1 pu per phase. The dc input current of the inverter is adjusted such that i_{w1} is rated.

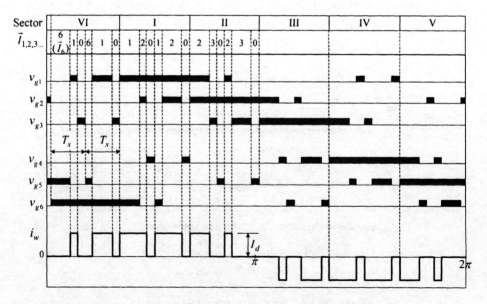

Figure 10.3-4 SVM switching sequence over a fundamental-frequency cycle.

10.3 Space Vector Modulation

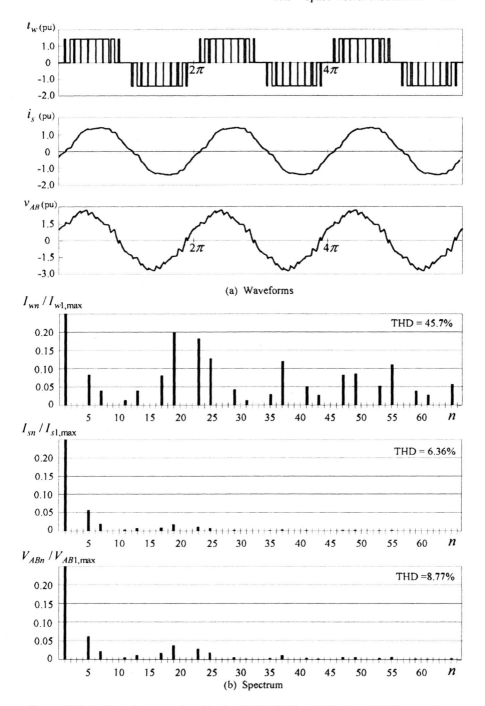

Figure 10.3-5 Waveforms produced by the SVM CSI ($f_1 = 60$ Hz, $f_{sw} = 540$ Hz, $m_a = 1$, and $C_f = 0.3$ pu).

208 Chapter 10 PWM Current Source Inverters

The spectrum for i_w, i_s, and v_{AB} are also shown in the figure, where I_{wn} is the rms value of the nth-order harmonic current in i_w and $I_{w,1,max}$ is the maximum rms fundamental-frequency current that can be found from (10.3-14) and (10.3-17):

$$I_{w1,max} = \frac{m_{a,max} \times I_d}{\sqrt{2}} = 0.707 I_d \quad \text{for } m_{a,max} = 1 \quad (10.3\text{-}20)$$

The PWM current i_w contains no even-order harmonics, and its THD is 45.7%. Similar to the two-level and three-level NPC inverters, the SVM current source in-

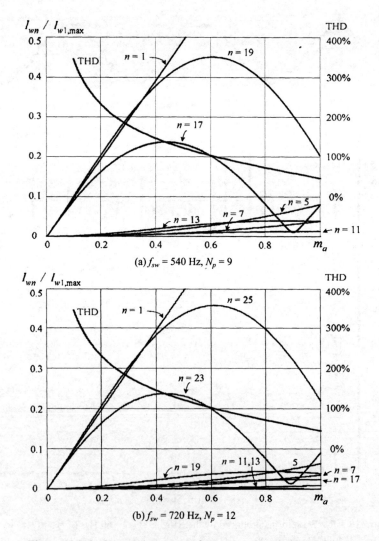

Figure 10.3-6 Harmonic content of i_w produced by the SVM CSI.

verter produces low-order harmonics, such as the 5th and 7th. The THD of i_s and v_{AB} is 6.36% and 8.77%, respectively.

10.3.5 Harmonic Content

The harmonic content of the PWM current i_w for the inverter operating at $f_1 = 60$ Hz with $f_{sw} = 540$ Hz and $f_{sw} = 720$ Hz is shown in Figs. 10.3-6a and 10.3-6b, respectively. There are a pair of dominant harmonics, whose order can be determined by $n = 2N_p \pm 1$. It is interesting to note that the THD curves for the two cases are almost identical. This is because the magnitudes of the two dominant harmonics in plot (a) are almost identical to those in plot (b).

10.3.6 SVM Versus TPWM and SHE

Table 10.3-2 provides a brief comparison among the three modulation schemes for the current source inverter. The main feature of the SVM scheme is fast dynamic response. This is in view of the fact that (a) its modulation index can be adjusted within a sampling period T_s and (b) the inverter PWM current i_w can be directly controlled by the bypass operation instead of dc current adjustment by the rectifier. Therefore, the SVM scheme is suitable for applications where a fast dynamic response is required. The SVM scheme has the lowest dc current utilization due to its bypass operation. The SHE scheme features superior harmonic performance. Its dynamic performance can also be improved by allowing dc current bypass operation for the quick adjustment of i_w. The performance of the TPWM modulation is somewhere between the SVM and SHE schemes.

10.4 PARALLEL CURRENT SOURCE INVERTERS

10.4.1 Inverter Topology

To increase the power rating of a CSI fed drive, two or more current source inverters can operate in a parallel manner [9, 10]. Figure 10.4-1 shows such a configura-

Table 10.3-2 Comparison of CSI Modulation Schemes

Item	SVM	TPWM	SHE
DC current utilization $I_{w1,max}/I_d$	0.707	0.74	0.73–0.78
Dynamic performance	High	Medium	Low
Digital implementation	Real time	Real time or look-up table	Look-up table
Harmonic performance	Adequate	Good	Best
DC current bypass operation	Yes	No	Optional

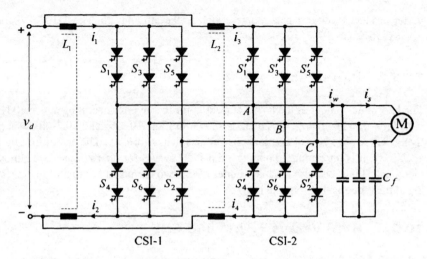

Figure 10.4-1 Parallel current source inverters for high-power MV drives.

tion where two inverters are connected in parallel. Each inverter has its own dc choke, but the two inverters share a common filter capacitor C_f at their output.

In practice, the parallel operation of the two inverters may cause unbalanced dc currents. The main causes for the unbalanced operation include (a) unequal on-state voltages of the semiconductor devices, which affects dc current balance in steady state; (b) variations in time delay of the gating signals of the two inverters, which affects both transient and steady-state current balance; and (c) manufacturing tolerance in dc choke parameters. In what follows, a space vector modulation algorithm is introduced, which can effectively solve the dc current unbalance problem [9].

10.4.2 Space Vector Modulation for Parallel Inverters

Following the procedure presented in Section 10.3, a space vector diagram composed of 19 current space vectors for the parallel inverters is illustrated in Fig. 10.4-2. These vectors can be divided into four groups according to their length: large, medium, small, and zero vectors. The 19 vectors correspond to 51 switching states given in Table 10.4-1. Each switching state is represented by four digits separated by a semicolon, the first two representing two on-state switches in CSI-1 and the last two denoting two on-state devices in CSI-2, respectively. For instance, switching state [12;16] for medium vector $\vec{I}_7$ indicates that switches S_1, S_2, S_1', and S_6' in the two inverters are turned on.

When designing the SVM algorithm for the parallel inverters, the effect of current vectors on the dc currents should be taken into account.

- Small and zero current vectors are not allowed to use since they introduce a bypass operation (shoot through) by turning on the two devices in the same

10.4 Parallel Current Source Inverters

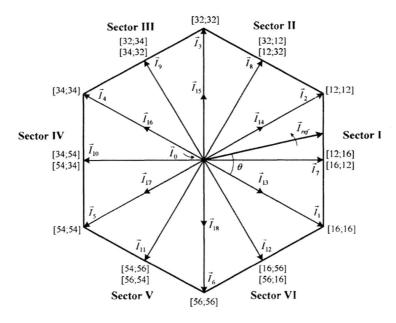

Figure 10.4-2 Space vector diagram for the parallel current source inverters.

Table 10.4-1 Classification of Space Vectors and Their Switching States

Classification	Current Vector	Switching State
Large vectors	$\vec{I}_1$	[16;16]
	$\vec{I}_2$	[12;12]
	$\vec{I}_3$	[32;32]
	$\vec{I}_4$	[34;34]
	$\vec{I}_5$	[54;54]
	$\vec{I}_6$	[56;56]
Medium vectors	$\vec{I}_7$	[12;16], [16;12]
	$\vec{I}_8$	[32;12], [12;32]
	$\vec{I}_9$	[32;34], [34;32]
	$\vec{I}_{10}$	[34;54], [54;34]
	$\vec{I}_{11}$	[54;56], [56;54]
	$\vec{I}_{12}$	[16;56], [56;16]
Small vectors	$\vec{I}_{13}$	[16;14], [14;16], [16;36], [36;16], [56;12], [12;56]
	$\vec{I}_{14}$	[12;14], [14;12], [12;52], [52;12], [16;32], [32;16]
	$\vec{I}_{15}$	[32;36], [36;32], [32;52], [52;32], [12;34], [34;12]
	$\vec{I}_{16}$	[34;36], [36;34], [34;14], [14;34], [32;54], [54;32]
	$\vec{I}_{17}$	[54;14], [14;54], [54;52], [52;54], [34;56], [56;34]
	$\vec{I}_{18}$	[56;52], [52;56], [56;36], [36;56], [54;16], [16;54]
Zero vector	$\vec{I}_0$	[14;14], [14;36], [14;52], [36;14], [36;52], [36;36], [52;14], [52;36], [52;52], [12;54], [54;12], [32;56], [56;32] [34;16], [16;34]

inverter leg simultaneously, resulting in increased switching frequency and energy loss. More importantly, the inverter output current in a practical MV drive is normally adjusted by the dc current instead of bypass operation through the modulation index control.

- Large vectors cannot be used for the dc current balance control. They turn on the devices in the same switch position of the two inverters. For instance, the switching state of $\vec{I}_1$ turns on S_1, S_6, S'_1, and S'_6 simultaneously, which does not have an effect on the dc currents.

- Only medium vectors can be utilized for the dc current balance control. Making use of the redundant switching states of the medium vectors, the dc currents in the two inverters can be controlled independently. The detailed analysis is given in the following section.

10.4.3 Effect of Medium Vectors on dc Currents

Figure 10.4-3 shows dc current paths in the parallel inverters with switching state [16;56] for medium vector $\vec{I}_{12}$. The dc current i_1 flows through S_1 and the load (phases A and B) and then flows back to the dc source through S_6 and S'_6. The dc current i_3 flows through S'_5 and the load (phases C and B) and then flows back to the dc source also through S_6 and S'_6. Assuming that the two inverters are identical and symmetrical, the dc currents in the negative dc buses are balanced, leading to $i_2 = i_4$.

The currents in the positive dc buses, i_1 and i_3, are affected by the load voltages. Assuming that the load phase voltage v_{AO} happens to be equal to v_{CO}, the two positive dc bus currents in this special case are balanced ($i_1 = i_3$). However, when v_{AO} is

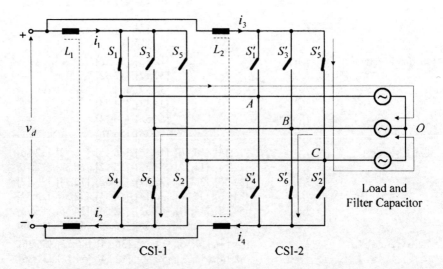

Figure 10.4-3 Current paths in the parallel inverters with switching state [16; 56].

Table 10.4-2 Effect of the Switching States of Medium Vector $\vec{I}_{12}$ on dc Currents

Switching State	Load Voltage	i_1	i_2	i_3	i_4
[16;56] and [56;16]	$v_{AO} = v_{CO}$	×	×	×	×
[16;56]	$v_{AO} > v_{CO}$	↓	×	↑	×
	$v_{AO} < v_{CO}$	↑	×	↓	×
[56;16]	$v_{AO} > v_{CO}$	↑	×	↓	×
	$v_{AO} < v_{CO}$	↓	×	↑	×

Symbol '×': dc currents not affected.

greater than v_{CO} ($v_{AO} > v_{CO}$), i_1 will decrease and in the meanwhile i_3 will increase. If $v_{AO} < v_{CO}$, a reverse action will take place for the dc currents.

Let's now consider the other switching state [56:16] of the medium vector $\vec{I}_{12}$. The effect of this switching state on i_1 and i_3 is exactly opposite to that of [16;56]. Table 10.4-2 provides a summary for both cases. It can be observed that for a given load voltage (except for $v_{AO} = v_{CO}$), one switching state can make the dc current increase while the other can make the same dc current decrease.

It should be further noted that the medium vectors in the even sectors (II, IV, and VI) of the space vector diagram can be used to adjust the positive dc bus currents (i_1 and i_3), but they do not have an effect on the negative dc bus currents. On the contrary, the medium vectors in the odd sectors (I, III, and V) can be used to control the negative dc bus currents (i_2 and i_4), but they do not affect the positive dc bus currents. Therefore, the positive and negative dc bus currents can be independently controlled by the two switching states of medium vectors.

10.4.4 dc Current Balance Control

To ensure a balanced operation for the two inverters, all the dc currents should be detected and controlled. The error signals for the detected dc currents are defined by

$$\Delta i_p = i_1 - i_3$$
$$\Delta i_n = i_2 - i_4 \quad (10.4\text{-}1)$$

where Δi_p and Δi_n are the current differences in the positive and negative dc buses, which should be zero when the inverters operate under the balanced condition. The error signals are then sent to two PI controllers for the dc current balance control. The output of the PI controllers are used to adjust the dwell time of medium vectors, given by

$$t_p = K\Delta i_p + \frac{1}{\tau}\int \Delta i_p dt$$
$$t_n = K\Delta i_n + \frac{1}{\tau}\int \Delta i_n dt \quad (10.4\text{-}2)$$

where t_p and t_n are the dwell times for the medium vectors in the even and odd sectors for the adjustment of positive and negative dc bus currents, and K and τ are the gain and time constant of the PI controllers, respectively.

Assuming that the reference vector $\vec{I}_{ref}$ is in sector I as shown in Fig. 10.4-2, $\vec{I}_{ref}$ can be synthesized by two large vectors ($\vec{I}_1$ and $\vec{I}_2$) and a medium vector ($\vec{I}_7$), that is,

$$\vec{I}_{ref} T_s = T_1 \vec{I}_1 + T_2 \vec{I}_2 + T_m \vec{I}_7 \qquad (10.4\text{-}3)$$

where T_1, T_2, and T_m are the dwell times for $\vec{I}_1$, $\vec{I}_2$, and $\vec{I}_7$, and T_s is the sampling period, respectively. To balance the dc currents, the dwell time T_m for the medium vector is adjusted by the output of the PI regulators:

$$T_m = \begin{cases} t_p & \text{for } \vec{I}_{ref} \text{ in sectors II, IV, and VI} \\ t_n & \text{for } \vec{I}_{ref} \text{ in sectors I, III, and V} \end{cases} \qquad (10.4\text{-}4)$$

The dwell times for the large vectors can be calculated by

$$T_1 = \frac{\sqrt{3} - \tan\theta}{\sqrt{3} + \tan\theta} T_s - \frac{1}{2} T_m$$
$$T_2 = T_s - T_1 - T_m \qquad (10.4\text{-}5)$$

where θ is the phase displacement between $\vec{I}_1$ and $\vec{I}_{ref}$ as shown in Fig. 10.4-2.

It should be pointed out that the three-phase load voltage should also be detected for the proper selection of the switching state of medium vectors. For the CSI drives, the combined power factor of the motor and the filter capacitor may vary from inductive to capacitive when the motor operates under different conditions. But this will not affect the dc current balance control since the switching states of the medium vectors are selected according to the sign of the measured load voltages, independent of the load power factor.

10.4.5 Experimental Verification

The SVM-based dc current balance control algorithm is implemented on a laboratory 5-hp (4-pole) induction motor drive using parallel current source inverters. The drive system operates under a light load condition with a maximum switching frequency of 420 Hz. To make the two inverters unbalanced on purpose, two power diodes are added to the dc circuit of CSI-2, one in the positive dc bus and the other in the negative bus. It is the diode voltage drop that makes the dc currents of the two inverters unbalanced.

Figure 10.4-4 shows the measured dc current waveforms during motor speed acceleration from 90 rpm to 1500 rpm. Without the dc current balance control, the current i_1 in CSI-1 is always higher than i_3 in CSI-2 as shown in oscillogram (a). When the motor operates at 1500 rpm with an increased dc current, the voltage drop on the diodes increases, resulting in a higher dc current difference. With the current balance control activated, both currents are kept almost the same as illustrated in oscillogram (b) during transient and steady-state operations.

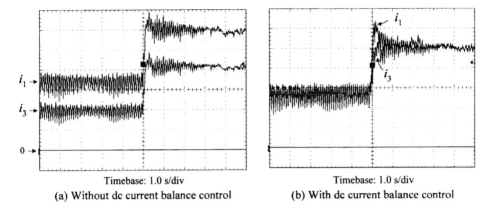

Timebase: 1.0 s/div
(a) Without dc current balance control

Timebase: 1.0 s/div
(b) With dc current balance control

Figure 10.4-4 Measured dc current waveforms during motor speed acceleration from 90 rpm to 1500 rpm.

Figure 10.4-5 shows the measured waveforms when the drive is running at 1500 rpm. The dc currents i_1 and i_3 in oscillogram (a) are kept balanced by the drive controller except for the middle portion of the current waveforms where the current balance control is temporarily disabled. The steady-state motor voltage and current waveforms are shown in oscillogram (b), which are close to sinusoidal.

10.5 LOAD-COMMUTATED INVERTER (LCI)

One of the well-known current source inverter topologies is the load-commutated inverter [11]. Figure 10.5-1 shows the typical LCI configuration for synchronous motor (SM) drives. On the dc side of the inverter, a dc choke L_d is required to provide a smooth dc current I_d. The inverter employs the SCR thyristor as a switching device instead of the symmetrical GCT for the PWM CSI. The SCRs do not have

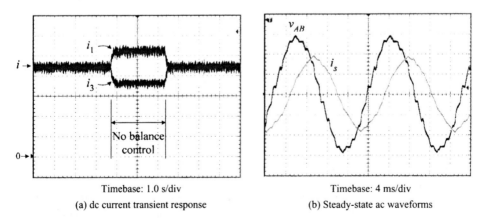

Timebase: 1.0 s/div
(a) dc current transient response

Timebase: 4 ms/div
(b) Steady-state ac waveforms

Figure 10.4-5 Measured inverter dc- and ac-side waveforms. at the motor speed of 1500 rpm.

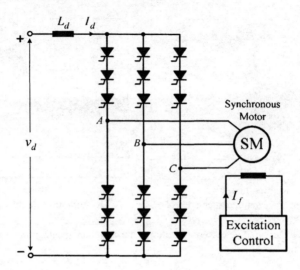

Figure 10.5-1 Load-commutated inverter for synchronous motor drive.

self-turn-off capability, but they can be naturally commutated by the load voltage with a leading power factor. The ideal load for the LCI is, therefore, synchronous motors operating at a leading power factor which can be easily achieved by adjusting the excitation current I_f.

The natural commutation of the SCRs is essentially accomplished by a leading EMF induced by the motor operating at certain speeds. At low motor speeds (typically lower than 10% the rated speed), the induced EMF may be too low to commutate the SCRs. In this case, the commutation is usually assisted by the front-end SCR rectifier.

The LCI-fed motor drive features low manufacturing cost and high efficiency mainly due to the use of low-cost SCR devices and lack of PWM operation. The LCI is a popular solution for very large drives, where the initial investment and operating efficiency are of great importance. A typical example is a 100 MW wind tunnel synchronous motor drive [4], where the efficiency of the power converters including the rectifier and inverter reaches 99%.

The main drawback of the LCI drive is its limited dynamic performance. However, the majority of the LCI drives are for fans, pumps, compressors, and conveyors, where the dynamic response is usually not a critical requirement. In addition, the power losses in the motor are high due to the large amount of harmonics in the stator current [12].

10.6 SUMMARY

This chapter focuses on the PWM current source inverter technologies for high-power medium-voltage drives. The operating principle for the CSI inverters is discussed. Three modulation techniques for the CSI are analyzed, which include trapezoidal pulse-width modulation (TPWM), selective harmonic elimination (SHE), and space vector modulation (SVM). These modulation techniques are developed for the high-power GCT inverters where the switching frequency of the inverter is normally below 500 Hz. This chapter also presents a new CSI topology using parallel inverters. An SVM-based dc current balance control algorithm is developed for the parallel inverters.

The PWM current source inverter features simple converter topology, motor-friendly waveforms, and reliable short-circuit protection, and therefore it is one of the popular converter topologies in the MV drive. Prior to the advent of GCTs, the GTO thyristors were dominant for the CSI drives. Although there are still a large number of installed GTO CSI drives in the field, this technology has been replaced by the GCT-based current source drives since the late 1990s.

REFERENCES

1. M. Hombu, S. Ueda, and A. Ueda, A Current Source GTO Inverter with Sinusoidal Inputs and Outputs, *IEEE Transactions on Industry Applications,* Vol. 23, No. 2, pp. 247–255, 1987.
2. P. Espelage, J. M. Nowak, and L. H. Walker, Symmetrical GTO Current Source Inverter for Wide Speed Range Control of 2300 to 4160 Volts, 350 to 7000 HP Induction Motors, *IEEE Industry Applications Society Conference (IAS),* pp. 302–307, 1988.
3. N. R. Zargari, S. C. Rizzo, et al., A New Current-Source Converter Using a Symmetric Gate-Commutated Thyristor (SGCT), *IEEE Transactions on Industry Applications,* Vol. 37, No. 3, pp. 896–903, 2001.
4. R. Bhatia, H. U. Krattiger, A. Bonanini, et al., Adjustable Speed Drive with a Single 100-MW Synchronous Motor, *ABB Review,* No. 6, pp. 14–20, 1998.
5. B. Wu, S. Dewan and G. Slemon, PWM-CSI Inverter Induction Motor Drives, *IEEE Transactions on Industry Applications,* Vol. 28, No. 1, pp. 64–71, 1992.
6. B. Wu, Pulse Width Modulated Current Source Inverter (CSI) Induction Motor Drives, Thesis, Master's in Applied Science, University of Toronto, 1989.
7. J. Wiseman, B. Wu and G. S. P. Castle, A PWM Current Source Rectifier with Active Damping For High Power Medium Voltage Applications, *IEEE Power Electronics Specialist Conference (PESC),* pp. 1930–1934, 2002.
8. J. Ma, B. Wu, and S. Rizzo, A Space Vector Modulated CSI-based ac Drive for Multimotor Applications, *IEEE Transactions on Power Electronics,* Vol. 16, No. 4, pp. 535–544, 2001.
9. D. Xu, N. Zargari, B. Wu, et al., A Medium Voltage AC Drive with Parallel Current Source Inverters for High Power Applications, *IEEE Power Electronics Specialist Conference (PESC),* 2005.
10. M. C. Chandorkar, D. M. Divan, and R. H. Lasseter, Control Techniques for Multiple Current Source GTO Converters, *IEEE Transactions on Industry Applications,* Vol. 31, No. 1, pp. 134–140, 1995.
11. J. P. McSharry, P. S. Hamer, et al., Design, Fabrication, Back-to-back Test of 14,200HP Two-Pole Cylindrical-Rotor Synchronous Motor for ASD Applications, *IEEE Transactions on Industry Applications,* Vol. 34, No. 3, pp. 526–533, 1998.
12. R. Emery and J. Eugene, Harmonic Losses in LCI-Fed Synchronous Motors, *IEEE Transactions on Industry Applications,* Vol. 38, No. 4, pp. 948–954, 2002.

APPENDIX

SHE SWITCHING ANGLES FOR INVERTER CIRCUIT OF FIG. 10.2-1

Harmonics Eliminated	Switching Angles			Harmonics Eliminated	Switching Angles			
	θ_1	θ_2	θ_3		θ_1	θ_2	θ_3	θ_4
5	18.00	—	—	7, 11, 17	11.70	14.12	24.17	—
7	21.43	—	—	7, 13, 17	12.69	14.97	24.16	—
11	24.55	—	—	7, 13, 19	13.49	15.94	24.53	—
13	25.38	—	—	11, 13, 17	14.55	15.97	25.06	—
5, 7	7.93	13.75	—	11, 13, 19	15.24	16.71	25.32	—
5, 11	12.96	19.14	—	13, 17, 19	17.08	18.23	25.84	—
5, 13	14.48	21.12	—	13, 17, 23	18.03	19.22	26.16	—
7, 11	15.23	19.37	—	5, 7, 11, 13	0.00	1.60	15.14	20.26
7, 13	16.58	20.79	—	5, 7, 11, 17	0.07	2.63	16.57	21.80
7, 17	18.49	23.08	—	5, 7, 11, 19	1.11	4.01	18.26	23.60
11, 13	19.00	21.74	—	5, 7, 13, 17	1.50	4.14	16.40	21.12
11, 17	20.51	23.14	—	5, 7, 13, 19	2.56	5.57	17.82	22.33
11, 19	21.10	23.75	—	5, 7, 17, 19	4.59	7.96	17.17	20.55
13, 17	21.19	23.45	—	5, 11, 13, 17	4.16	6.07	16.79	22.04
13, 19	21.71	23.94	—	5, 11, 13, 19	5.13	7.26	17.57	22.72
5, 7, 11	2.24	5.60	21.26	5, 11, 17, 19	6.93	9.15	17.85	22.77
5, 7, 13	4.21	8.04	22.45	5, 13, 17, 19	7.80	9.82	18.01	23.25
5, 7, 17	6.91	11.96	25.57	7, 11, 13, 17	5.42	6.65	18.03	22.17
5, 11, 13	7.81	11.03	22.13	7, 11, 13, 19	6.35	7.69	18.67	22.74
5, 11, 17	10.16	14.02	23.34	7, 11, 17, 19	8.07	9.44	19.09	22.93
5, 13, 17	11.24	14.92	22.98	7, 13, 17, 19	8.88	10.12	19.35	23.22
7, 11, 13	9.51	11.64	23.27	11, 13, 17, 19	10.39	11.14	20.56	23.60

Chapter 11

PWM Current Source Rectifiers

11.1 INTRODUCTION

With the advent of gate-commutated thyristor (GCT) devices in the late 1990s, the PWM current source rectifier using symmetrical GCT switches is increasingly used in current source fed medium-voltage (MV) drives. Compared with the multipulse SCR rectifiers presented in Chapter 4, the PWM rectifier features improved input power factor, reduced line current distortion, and superior dynamic response.

The PWM current source rectifier (CSR) normally requires a three-phase filter capacitor at its input. The capacitor provides two basic functions: (a) to assist the commutation of switching devices and (b) to filter out line current harmonics. However, the use of the filter capacitor may cause LC resonances and affect the input power factor of the rectifier as well.

This chapter addresses four main issues of the current source rectifiers, including converter topologies, PWM schemes, power factor control, and active damping control for LC resonance suppression. The important concepts are elaborated with simulations and experiments.

11.2 SINGLE-BRIDGE CURRENT SOURCE RECTIFIER

11.2.1 Introduction

Figure 11.2-1 shows the circuit diagram of a single-bridge GCT current source rectifier [1–3]. When the rectifier is used in high-power MV drives as a front end, two or more GCTs can be connected in series. The **line inductance** L_s on the ac side of the rectifier represents the total inductance between the utility supply and the rectifier, including the equivalent inductance of the supply, leakage inductances of isolation transformer if any, and inductance of the line reactor for line current THD reduction. The line inductance L_s is normally in the range of 0.1 to 0.15 per unit (pu).

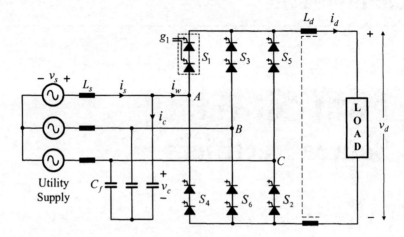

Figure 11.2-1 Single-bridge GCT current source rectifier.

The definition of per unit system is given in Chapter 3.

The PWM rectifier requires a **filter capacitor** C_f to assist the commutation of the GCT devices and filter out harmonic currents. The capacitor size is dependent on a number of factors such as the rectifier switching frequency, LC resonant mode, required line current THD, and input power factor. It is normally in the range of 0.3 to 0.6 pu for high-power PWM rectifiers with a switching frequency of a few hundred hertz.

On the dc side of the rectifier, a **dc choke** L_d is required to smooth the dc current. The choke usually has a magnetic core with two coils, one connected in the positive dc bus and the other in the negative bus. Such an arrangement is preferred in practice for motor common-mode voltage reduction [4]. To limit the dc link current ripple to an acceptable level (< 15%), the size of the dc choke is normally in the range of 0.5 to 0.8 pu.

11.2.2 Selective Harmonic Elimination

As discussed in Chapter 10, the **selective harmonic elimination** (SHE) is considered as an optimum modulation scheme, which provides a superior harmonic profile with a minimum switching frequency. Unlike the SHE modulation schemes for current source inverters where the modulation index is usually fixed to its maximum value, the SHE scheme for the CSR should provide an adjustable modulation index for dc current control in addition to the harmonic elimination [5, 6].

Figure 11.2-2 shows a typical half-cycle waveform of the rectifier PWM current i_w. The current waveform is designed such that the **switching constraints** for the current source converters discussed in the previous chapter are satisfied. There are six current pulses per half-cycle of the fundamental frequency with only three independent **switching angles**, β_1, β_2, and β_0. These angles can be used to eliminate

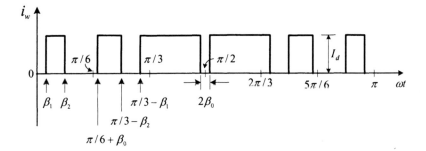

Figure 11.2-2 PWM current waveform (half-cycle) with three independent switching angles (β_1, β_2, and β_0).

two harmonics and in the meanwhile provide an adjustable modulation index.

The PWM current waveform in Fig. 11.2-2 can be expressed in Fourier series as

$$i_w(\omega t) = \sum_{n=1}^{\infty} a_n \sin(n\omega t) \qquad (11.2\text{-}1)$$

where

$$a_n = \frac{4}{\pi} \int_0^{\frac{\pi}{2}} i_w(\omega t)\sin(n\omega t)d(n\omega t)$$

$$= \frac{4I_d}{n\pi}\left\{\cos(n\beta_1) - \cos(n\beta_2) + \cos\left(n\left(\frac{\pi}{6}+\beta_0\right)\right) - \cos\left(n\left(\frac{\pi}{3}-\beta_2\right)\right)\right.$$

$$\left. + \cos\left(n\left(\frac{\pi}{3}-\beta_1\right)\right) - \cos\left(n\left(\frac{\pi}{2}-\beta_0\right)\right)\right\} \qquad (11.2\text{-}2)$$

To eliminate two dominant harmonics such as the 5th and 7th, we have

$$F_1 = \cos(5\beta_1) - \cos(5\beta_2) + \cos\left(5\left(\frac{\pi}{6}+\beta_0\right)\right) - \cos\left(5\left(\frac{\pi}{3}-\beta_2\right)\right)$$

$$+ \cos\left(5\left(\frac{\pi}{3}-\beta_1\right)\right) - \cos\left(5\left(\frac{\pi}{2}-\beta_0\right)\right) = 0 \qquad (11.2\text{-}3)$$

$$F_2 = \cos(7\beta_1) - \cos(7\beta_2) + \cos\left(7\left(\frac{\pi}{6}+\beta_0\right)\right) - \cos\left(7\left(\frac{\pi}{3}-\beta_2\right)\right)$$

$$+ \cos\left(7\left(\frac{\pi}{3}-\beta_1\right)\right) - \cos\left(7\left(\frac{\pi}{2}-\beta_0\right)\right) = 0 \qquad (11.2\text{-}4)$$

A third equation can be derived for modulation index adjustment

$$F_3 = \frac{a_1}{I_d} - m_a = \frac{4}{\pi}\left\{\cos\beta_1 - \cos\beta_2 + \cos\left(\frac{\pi}{6} + \beta_0\right) - \cos\left(\frac{\pi}{3} - \beta_2\right)\right.$$

$$\left. + \left(\frac{\pi}{3} - \beta_1\right) - \left(\frac{\pi}{2} - \beta_0\right)\right\} - m_a = 0 \qquad (11.2\text{-}5)$$

where m_a is **amplitude modulation index**, given by

$$m_a = \frac{\hat{I}_{w1}}{I_d} \qquad (11.2\text{-}6)$$

$\hat{I}_{w1}$ in (11.2-6) is the peak fundamental-frequency component of the PWM current and I_d is the average dc current. Using Newton–Raphson iteration algorithm introduced in Chapter 10 or other numerical methods to solve (11.2-3) to (11.2-5) simultaneously, we can obtain all the switching angles at various modulation indices. The calculated results are given in Table 11.2-1. The maximum modulation index is 1.03, at which β_0 becomes zero and the notch in the centre of the PWM waveform disappears. The switching angles β_1 and β_2 at $m_a = 1.03$ are identical to those for the current source inverter with 5th and 7th harmonic elimination.

For a given value of the switching angels β_1, β_2, and β_0, the gate signals for the switching devices in the CSR can be arranged. An example is shown in Fig. 11.2-3, where v_{g1} and v_{g4} are the gate signals for S_1 and S_4 in rectifier leg A, respectively. The gate signal v_{g1} is composed of six pulses, of which one is the **bypass pulse**, defined by θ_{11} and θ_{12}. The notch $2\beta_0$ in the center of the i_w waveform is realized by turning on S_1 and S_4 simultaneously. The dc current i_d is then bypassed (shorted) by the rectifier, leading to $i_w = 0$. During the bypass interval, i_d is kept constant by the large dc choke.

Based on the gate signals in Fig. 11.2-3, the **switching frequency** of the GCT device can be calculated by $f_{sw} = f_s \times N_p = 60 \text{ Hz} \times 6 = 360 \text{ Hz}$, where f_s is the supply frequency and N_p is the number of pulses per cycle of the supply frequency. This is the lowest possible switching frequency that can be used to eliminate two harmonics and in the meantime to provide an adjustable modulation index. All the **gating angles**, θ_1 to θ_{12}, can be calculated from Table 11.2-1 and are plotted in Fig. 11.2-4. Their numerical values are given in the Appendix of this chapter.

Table 11.2-1 Switching Angles for the Single-Bridge CSR with 5th and 7th Harmonic Elimination

m_a:	0.1	0.2	0.3	0.4	0.5	0.6	0.7	0.8	0.9	1.0	1.03
β_1:	−13.5°	−11.9°	−10.3°	−8.60°	−6.86°	−5.00°	−3.98°	−0.67°	2.17°	6.24°	7.93°
β_2:	14.2°	13.5°	12.7°	12.0°	11.4°	10.8°	10.4°	10.3°	10.8°	12.6°	13.8°
β_0:	13.6°	12.2°	10.9°	9.5°	8.0°	6.6°	5.1°	3.6°	2.1°	0.5°	0

11.2 Single-Bridge Current Source Rectifier

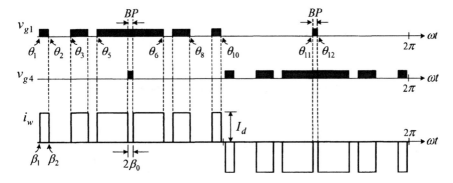

Figure 11.2-3 Gate signal arrangement for the single-bridge current source rectifier ($m_a = 0.9$).

It is interesting to note that when the modulation index m_a is lower than 0.826, gating angle θ_1 becomes negative and θ_{10} is larger than 180°. Figure 11.2-5 shows such a case with $m_a = 0.4$, $\theta_1 = -8.6°$, and $\theta_{10} = 188.6°$. There are four bypass intervals ($BP1$ to $BP4$) per cycle. The bypass intervals, $BP2$ and $BP4$, are created by bypass pulses while the other two, $BP1$ and $BP3$, are due to the overlapping of the gate signals.

The harmonic content of the rectifier PWM current i_w is illustrated in Fig. 11.2-6, where I_{w1} and I_{wn} are the rms values of the fundamental-frequency and nth-order

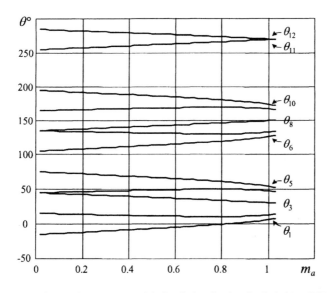

Figure 11.2-4 Gating angles versus modulation index for the single-bridge CSR with 5th and 7th harmonic elimination.

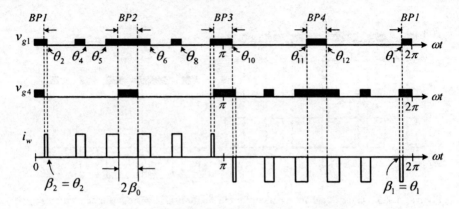

Figure 11.2-5 Four bypass intervals ($m_a < 0.826$).

harmonic currents, respectively. The fundamental current I_{w1} increases linearly with m_a. With the 5th and 7th harmonics eliminated, the magnitudes of remaining harmonics are high, especially for the 11th harmonic. However, these harmonic currents can be substantially attenuated by the filter capacitor C_f and line inductance L_s.

Figure 11.2-7 shows simulated current and voltage waveforms of the single-bridge rectifier with 5th and 7th harmonic elimination. The rectifier has a line inductance of 0.1pu and filter capacitor of 0.6 pu. It operates at the rated dc current with a modulation index of 0.9. As shown in the figure, the THD of the PWM cur-

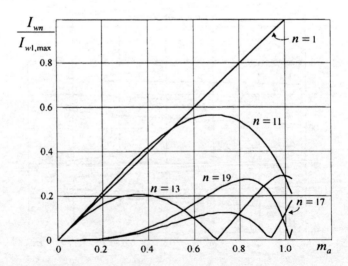

Figure 11.2-6 Harmonic content of I_w in the single-bridge CSR with 5th and 7th harmonic elimination.

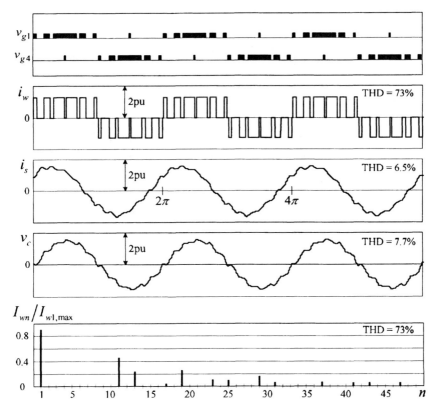

Figure 11.2-7 Simulated waveforms in the single-bridge rectifier ($L_s = 0.1$ pu, $C_f = 0.6$ pu, and $m_a = 0.9$).

rent i_w is quite high (73%) whereas the THD of the line (source) current i_s is only 6.5% due to the use of SHE modulation and the LC filter.

Figure 11.2-8 shows the current waveforms measured from a laboratory CSR system operating under the condition of $f_s = 60$ Hz, $m_a = 0.7$, $N_p = 6$, $L_s = 0.1$ pu and $C_f = 0.66$ pu. The traces from the top to bottom are the switch S_1 current i_{GCT}, the rectifier PWM current i_w and the line current i_s. With the modulation index lower than 0.826, the dc current i_d should be bypassed by the rectifier four times per cycle of the supply frequency. The bypasses are reflected in the differences between i_{GCT} and positive portion of i_w. The line current i_s is close to sinusoidal due to the SHE modulation and the filter capacitor.

11.2.3 Rectifier dc Output Voltage

The dc output voltage v_d of the rectifier can be adjusted by two methods, modulation index (m_a) control and delay angle (α) control, where α is the phase displacement between the utility supply voltage v_s and the rectifier fundamental current i_{w1}.

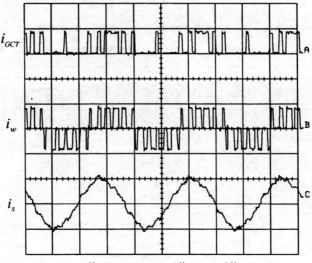

All traces: 0.75 pu/div, 5 ms/div.

Figure 11.2-8 Current waveforms measured from a laboratory single-bridge CSR ($f_s = 60$ Hz, $m_a = 0.7$, $L_s = 0.1$ pu, and $C_f = 0.66$ pu).

The operating principle of delay angle control is the same as that in phase-controlled SCR rectifiers.

The ac input power can be expressed as

$$P_{ac} = \sqrt{3} V_{LL} I_{w1} \cos \alpha \qquad (11.2\text{-}7)$$

where V_{LL} is the rms line-to-line voltage of the utility supply. The dc output power is given by

$$P_d = V_d I_d \qquad (11.2\text{-}8)$$

where V_d and I_d are the average dc voltage and current, respectively. Neglecting the power losses in the rectifier, the ac input power should be equal to the dc output power,

$$\sqrt{3} V_{LL} I_{w1} \cos \alpha = V_d I_d \qquad (11.2\text{-}9)$$

from which

$$V_d = \sqrt{3/2} \, V_{LL} m_a \cos \alpha \qquad (11.2\text{-}10)$$

where $m_a = \hat{I}_{w1}/I_d = \sqrt{2} I_{w1}/I_d$.

Considering the maximum modulation index of $m_{a,max} = 1.03$ for the single-bridge CSR, the maximum dc output voltage is

$$V_{d,max} = \sqrt{3/2}\, V_{LL} m_{a,max} \cos 0° = 1.26 V_{LL} \qquad \text{for } \alpha = 0 \qquad (11.2\text{-}11)$$

Compared with a phase-controlled SCR rectifier where the maximum dc output voltage is $1.35 V_{LL}$, the single-bridge PWM rectifier has a dc voltage derating of 7%.

11.2.4 Space Vector Modulation

The SHE modulation is an off-line PWM scheme. All the switching angles are precalculated and then stored in a look-up table for digital implementation. Although SHE scheme provides a superior harmonic performance with a minimum switching frequency, it may not be suitable for the current source rectifiers that require an instantaneous adjustment of modulation index.

The space vector modulation (SVM) is an on-line scheme suitable for real-time digital implementation. Compared with the SHE modulation, the SVM provides faster dynamic response and greater flexibility since the modulation index can be adjusted within each sampling period. As an example, the SVM modulation can be used in the active damping control, where the modulation index should be adjusted dynamically to suppress oscillations caused by LC resonances. The SVM scheme discussed in Chapter 10 for the CSI can be directly applied to the CSR, and therefore is not discussed in this chapter.

11.3 DUAL-BRIDGE CURRENT SOURCE RECTIFIER

11.3.1 Introduction

Figure 11.3-1 shows the converter configuration of a dual-bridge current source rectifier [7, 8]. This topology is composed of two identical single-bridge CSRs powered by a **phase shifting transformer** with two secondary windings, one connected in delta (Δ) and the other in wye (Y). The line-to-line voltage of each secondary winding is normally half of that of the primary winding. The total line inductance L_s is referred to the secondary side of the transformer for the convenience of discussion. The filter capacitor C_f is normally in the range of 0.15 to 0.3 pu, which is smaller than that for the single-bridge CSR.

The dual-bridge rectifier has the following features:

- **Sinusoidal Line Current.** The transformer is used to cancel the 5th, 7th, 17th, and 19th harmonic currents while the PWM technique is employed to eliminate the 11th and 13th harmonics. As a result, the primary-side line current i_A does not contain any harmonics lower than the 23rd. The other high-order harmonics can be attenuated by the filter capacitor C_f.

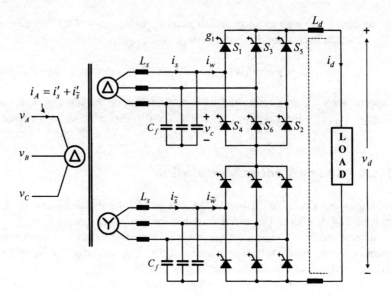

Figure 11.3-1 Dual-bridge GCT current source rectifier.

- **Low Switching Frequency.** The rectifier can draw sinusoidal input current with a low switching frequency, typically 360 Hz or 420 Hz for the elimination of the 11th and 13th harmonics.
- **Reliable Operation.** No GCT devices are connected in series in the dual-bridge CSR, which enhances the reliability of the system.
- **Suitable for Retrofit Applications.** The dual-bridge rectifier requires a phase shifting transformer for harmonic cancellation. The transformer can also block common-mode voltages that would otherwise appear on the motor windings, causing premature failure of the winding insulation system. The MV drive with the dual-bridge CSR as a front end is suited for retrofit applications, where a standard ac motor is usually employed.

11.3.2 PWM Schemes

To design the SHE switching pattern for the dual-bridge rectifier, the following requirements should be satisfied:

- To eliminate 11th and 13th harmonics in the PWM current of each bridge.
- To provide a full range of modulation index control.
- To minimize switching frequency.

In addition, the PWM pattern design must satisfy the **switching constraints** for current source converters: Only two switching devices should conduct at any time, one connected to the positive dc bus and the other to the negative dc bus.

The switching pattern developed for the single-bridge CSR with one bypass pulse per cycle (Fig. 11.2-3) does not work for the 11th and 13th harmonic elimination. Figure 11.3-2 shows two new switching patterns for the dual-bridge CSR. Pattern A is suited for the rectifier operating at low modulation indices. The gate signal v_{g1} for switch S_1 is composed of six pulses per cycle of the supply frequency, of which three are the bypass pulses specified by θ_7 to θ_{12}. The resultant device switching frequency is 360 Hz, six times the supply frequency of 60 Hz. Pattern B is developed for the rectifier operating at high modulation indices. It can be observed that S_1 switches seven times per cycle, resulting in a switching frequency of 420 Hz.

Figure 11.3-3 illustrates the calculated gating angles versus modulation index m_a for both patterns. Pattern A is valid for $0.02 \leq m_a \leq 0.857$. It is noted that the difference between the gating angles θ_9 and θ_{10} reduces with the increase of m_a. The two angles finally merge at $m_a = 0.857$, beyond which the 11th and 13th harmonics are no longer eliminated. Similarly, pattern B is valid only for $0.84 \leq m_a \leq 1.085$. These two patterns can then be combined to provide a continuous adjustment of m_a over the full operating range of the rectifier.

11.3.3 Harmonic Contents

The harmonic content in the rectifier PWM current i_w is shown in Fig. 11.3-4, where I_{w1} and I_{wn} are the rms values of the fundamental-frequency and nth-order harmonic currents, respectively. The fundamental current I_{w1} increases linearly with m_a. The current i_w contains the 5th, 7th, 17th, and 19th harmonics, but these harmonics can be eliminated by the phase shifting transformer (refer to Chapter 5 for details). It is interesting to note that the harmonic curves are smoothly joined for the two patterns.

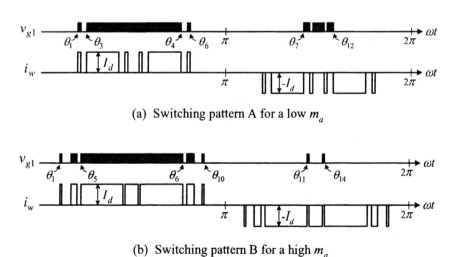

(a) Switching pattern A for a low m_a

(b) Switching pattern B for a high m_a

Figure 11.3-2 Two switching patterns for the dual-bridge CSR with 11th and 13th harmonic elimination.

230 Chapter 11 PWM Current Source Rectifiers

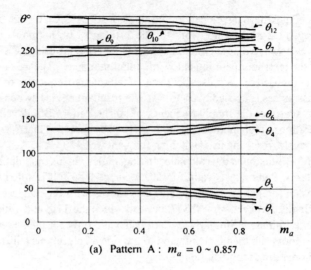

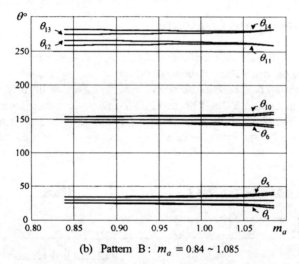

Figure 11.3-3 Gating angles θ versus modulation index m_a for the dual-bridge CSR with 11th and 13th harmonic elimination.

Figure 11.3-5 shows the current waveforms measured from a laboratory dual-bridge CSR system operating at $m_a = 0.5$ of pattern A and $m_a = 1.02$ of pattern B, respectively. The total line inductance L_s is 0.067 pu and the filter capacitance is 0.15 pu.

The traces from top to bottom are the transformer primary line current i_A, the secondary line current i_s, and the rectifier PWM current i_w. The waveform of the primary current i_A is virtually sinusoidal, which is the main feature of the dual-bridge rectifier.

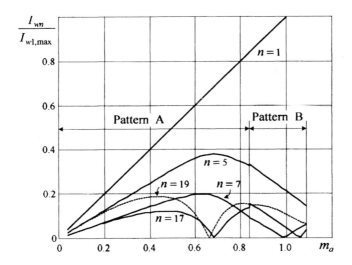

Figure 11.3-4 Harmonic content of i_w produced by combined PWM patterns A and B.

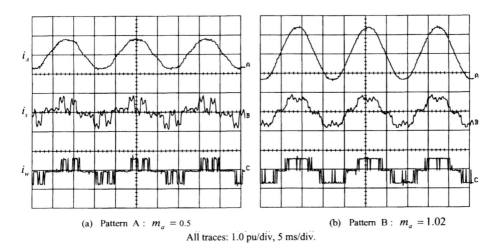

(a) Pattern A : $m_a = 0.5$ (b) Pattern B : $m_a = 1.02$
All traces: 1.0 pu/div, 5 ms/div.

Figure 11.3-5 Current waveforms measured from a laboratory dual-bridge CSR.

11.4 POWER FACTOR CONTROL

11.4.1 Introduction

The PWM current source rectifier requires a three-phase filter capacitor at its input terminals. Its capacitance is typically in a range of 0.3 to 0.6 pu for the single-bridge CSR with a switching frequency of 200 Hz. The capacitor size can be reduced when the CSR operates at higher switching frequencies or the dual-bridge rectifier is used. Neglecting a small voltage drop across the line inductance, the capacitor voltage is approximately equal to the supply voltage, resulting in a constant current flowing through the capacitor regardless of the operating condition of the rectifier.

The rectifier dc output current can be controlled by modulation index m_a. Alternatively, it can also be adjusted by delay angle α in the same manner as that for phase-controlled SCR rectifiers. The delay angle control produces a lagging power factor, which compensates the leading power factor caused by the filter capacitor. By controlling both modulation index and delay angle simultaneously, the rectifier can potentially achieve unity power factor operation while its dc current can also be controlled [9, 10].

11.4.2 Simultaneous α and m_a Control

Figure 11.4-1 shows two phasor diagrams for the single-bridge CSR, where all the voltage and current phasors, such as I_s, I_w, and V_c, represent their fundamental-frequency components only. The subscript "1" designated for the fundamental frequency is omitted for simplicity. The phasor diagram in part (b) illustrates the operation of the rectifier with modulation index control only. The PWM current I_w is in phase with the supply voltage V_s while the line current I_s leads V_s by the power factor angle ϕ. The input power factor of the rectifier is then given by

$$\text{PF} = \text{DF} \times \cos\phi = \cos\phi \qquad (11.4\text{-}1)$$

where the distortion power factor DF is assumed to be unity, which is based on the fact that the waveform of the line current i_s is close to sinusoidal. Under this assumption, the control of the power factor in this section is essentially to control the displacement power factor of the rectifier.

The input power factor can be improved by increasing the delay angle α between I_w and V_s as shown in Fig. 11.4-1c and in the meanwhile increasing the modulation index m_a to compensate the dc voltage reduction due to the increase of α. The dc voltage is a function of m_a and α, given by

$$V_d = \sqrt{3/2}\, V_{LL} m_a \cos\alpha \qquad (11.4\text{-}2)$$

To achieve a unity power factor operation, the delay angle should satisfy

$$\alpha \approx \sin^{-1}\frac{I_c}{I_w} \approx \sin^{-1}\frac{\omega C_f V_s}{m_a I_d} \qquad (11.4\text{-}3)$$

where the voltage drop on the line inductance is neglected.

However, the unity power factor operation is not always achievable. The phasor diagram composed of I'_w, I'_c, and I'_s in Fig. 11.4-1c illustrates such a case, where the rectifier operates under the light load condition. Obviously, the PWM current I'_w is too small to provide enough lagging current component to cancel the leading capacitor current. Therefore, the power factor control scheme should be designed such

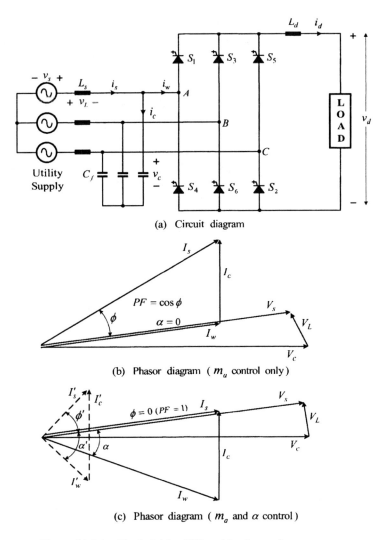

Figure 11.4-1 Single-bridge CSR and its phasor diagram.

that the rectifier will (a) operate at a unity power factor when it is achievable and (b) produce the highest possible power factor when the unity power factor operation is not achievable. The latter can be realized by

- increasing the PWM current I_w to its highest value by setting m_a to its maximum value such that I_w is able to produce the largest lagging current component; and then
- adjusting the delay angle α to produce a required dc current set by its reference.

Figure 11.4-2 shows a power factor control scheme derived from the above-discussed principle. There exist two control loops. In the m_a control loop, the supply voltage V_s and line current I_s are detected though a low-pass filter (LPF) and then sent to the power factor angle detector. The detected power factor angle ϕ is compared with its reference ϕ^*, which is normally set to zero, demanding a unity power factor operation. The resultant error signal $\Delta\phi$ is used to control modulation index m_a through a PI regulator.

In the α control loop, the detected dc current I_d is compared with its reference I_d^*. The error signal ΔI is then sent to a PI for delay angle control. The dc current is essentially controlled by both m_a and α. The PWM generator produces the gate signals for the GCTs in the CSR based on calculated modulation index m_a and delay angle α. The voltage zero crossing detector (VZD) provides a reference for the delay angle control.

The control scheme can guarantee that the input power factor of the rectifier is unity when it is achievable. Assuming that the line current I_s leads the supply voltage V_s by an angle ϕ due to a change in load, an error signal $\Delta\phi$ is generated. This error signal results in a higher modulation index m_a, which boosts the dc voltage V_d governed by (11.4-2). The increase in V_d makes the dc current I_d rise, which causes the α control loop to respond. The control loop tries to bring I_d back to the value set by I_d^* by increasing α. The increase in α causes a reduction in ϕ, which improves the input power factor. This process continues until the unity power factor is reached, at which the power factor angle ϕ equals zero, the phase displacement error $\Delta\phi$ equals zero, the dc current I_d equals I_d^*, and the rectifier operates at a new operating point.

When the rectifier operates under light load conditions, the unity power factor operation may not be achievable. Similar to the case discussed above, the modulation index m_a keeps increasing due to $\Delta\phi$, and in the meantime the delay angle α also keeps increasing for the dc current adjustment and power factor improvement.

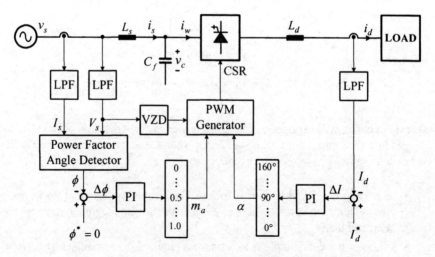

Figure 11.4-2 Block diagram of power factor control scheme.

Since $\Delta\phi$ will not be reduced to zero, the process continues until the PI regulator in the m_a loop is saturated, at which m_a reaches its maximum value $m_{a,\max}$ and the delay angle α also reaches a value that produces the highest possible input power factor while maintains I_d at its reference value.

Obviously, the transition between the two operation modes, the unity and maximum power factor operations, is smooth and seamless. No extra measures should be taken for the transition.

11.4.3 Power Factor Profile

Figure 11.4-3 shows the power factor profile for the rectifier operating at a dc voltage of 0.5 pu and filter capacitor of 0.4 pu. With simultaneous m_a and α control, the rectifier can operate with a unity power factor in Region A. When the unity power factor is not achievable under light load conditions, the controller can produce a highest possible power factor as shown in Region B. The power factor of the rectifier with the m_a control only is also plotted in the figure, where the power factor never reaches unity due to the leading capacitor current. It is clear that the combination of the m_a and α control significantly improves the input power factor of the rectifier.

Figure 11.4-4 shows the waveforms of dc current i_d, delay angle α, modulation index m_a, and PWM current i_w measured from a laboratory single-bridge CSR. The rectifier has a total line inductance of $L_s = 0.1$ pu, filter capacitance of $C_f = 0.6$ pu, and load resistance of 0.64 pu. It operates at a dc current of with a switching frequency of 360 Hz for the 5th and 7th harmonic elimination. Under this operating condition, the rectifier cannot achieve unity power factor operation since the PWM current i_w is not high enough to fully compensate for the leading capacitor current

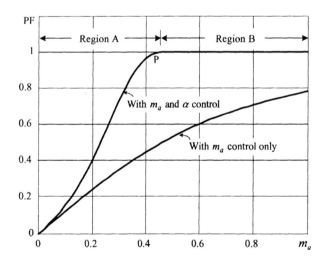

Figure 11.4-3 Rectifier power factor versus modulation index m_a ($V_d = 0.5$ pu and $C_f = 0.4$ pu).

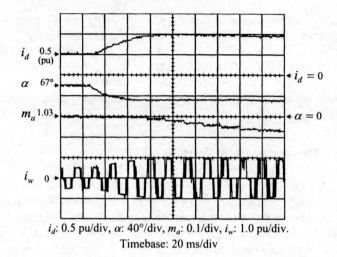

i_d: 0.5 pu/div, α: 40°/div, m_a: 0.1/div, i_w: 1.0 pu/div.
Timebase: 20 ms/div

Figure 11.4-4 Rectifier dynamic response to a step increase in I_d^*, leading to a unity power factor operation.

of $I_c = 0.6$ pu. The PI regulator in the m_a control loop is saturated, keeping m_a at its maximum value of 1.03. The delay angle α is adjusted by its PI regulator to 67°, at which the dc current I_d equals $I_d^* = 0.5$ pu and the power factor PF equals 0.93 (leading), which is the maximum achievable value.

When the load current is increased from 0.5 pu to 1.0 pu by a step increase in I_d^*, the rectifier is able to achieve unity power factor operation since the PWM current i_w is now sufficiently high to compensate the leading capacitor current. The load current i_d rises from 0.5 pu to 1.0 pu, the delay angle α decreases from its original value of 67° to 35°, and the modulation index m_a falls from its maximum value to 0.96, at which the rectifier operates at the unity power factor. Figure 11.4-5 shows the waveforms of the supply voltage v_s, line current i_s, and PWM current i_w when the rectifier reaches the unity power factor operation, where the line current i_s is in phase with the supply voltage v_s.

It is worth noting that the power factor control scheme does not require any system parameters, such as the values of the line inductance or filter capacitor. Variations in the line inductance due to the power system operation or the changes in filter capacitor size do not affect the process of tracking unity or maximum power factor, which is desirable in practice.

11.5 ACTIVE DAMPING CONTROL

11.5.1 Introduction

The filter capacitor C_f at the input of the CSR forms an LC resonant circuit with the total line inductance L_s of the system. The LC resonant mode may be excited by the

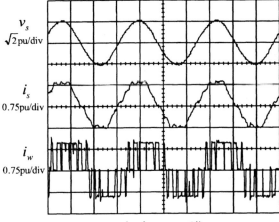

Figure 11.4-5 Waveforms measured from the single-bridge CSR at unity power factor operation.

harmonic voltages in the utility supply and harmonic currents produced by the rectifier. In this section, the principle of active damping is introduced, based on which a control algorithm is developed. The effectiveness of the active damping control is investigated by simulation and verified by experiments.

11.5.2 Series and Parallel Resonant Modes

The LC resonant circuit in the PWM rectifier has two modes of resonance depending on the source of excitation. Assuming a balanced three-phase system, the LC circuit can be analyzed on a per-phase basis. The **series resonant mode** can be obtained by looking into the LC circuit and rectifier of Fig. 11.4-1 from the utility supply, where the rectifier can be considered as a constant current source and thus can be open-circuited as shown in Fig. 11.5-1a. The line inductance L_s is essentially placed in series with the filter capacitor C_f. The LC circuit in high-power rectifiers

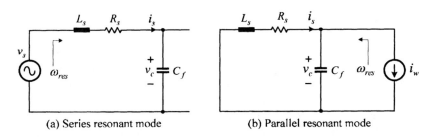

(a) Series resonant mode   (b) Parallel resonant mode

Figure 11.5-1 Series and parallel resonant modes in a current source rectifier.

is lightly damped due to a very low equivalent series resistance R_s, which is the total resistance of the distribution line, line reactor and isolation transformer if any. The series LC resonance may be excited by voltage harmonics in the utility supply.

The **parallel resonant mode** can be identified by looking into the LC circuit and utility supply from the rectifier, where the supply is essentially a constant voltage source and thus can be short-circuited as shown in Fig. 11.5-1b. The parallel resonance may be excited by the current harmonics generated by the PWM rectifier. Neglecting R_s, the resonant mode of the series and parallel resonances can be found from $\omega_{res} = 1/\sqrt{L_s C_f}$. With the filter capacitor C_f in the range of 0.3 to 0.6 pu and the total line inductance L_s between 0.1 and 0.15 pu, the resultant LC resonant frequency ω_{res} is in the range of 3.3 to 5.8 pu.

The LC resonance problem is traditionally addressed by selecting a proper LC resonant frequency using reasonable values for L_s and C_f such that the frequency of the LC resonance is lower than that of the lowest harmonic produced by the rectifier. However, the source inductance may vary with power system operations. It is possible for the source inductance to change such that the LC resonant mode is excited by a low-order harmonic. The use of the active damping control can effectively suppress the LC resonances, and simplifies the design of the LC filter as well.

11.5.3 Principle of Active Damping

It is well known that an LC resonance can be suppressed by adding a physical damping resistor to the resonant circuit. For the current source rectifier, the most effective location for the damping resistor R_p is in parallel with the filter capacitor C_f as shown in Fig. 11.5-2a. This location is effective at damping both series and parallel resonances. However, adding a resistor to the system results in additional power losses and thus is not a practical solution.

Active damping uses the rectifier to emulate a damping resistance R_p in the system [11–13]. Figure 11.5-2b illustrates the principle of one of the active damping schemes [11]. The damping current i_p can be generated by the PWM rectifier through the following steps:

- Detect the instantaneous capacitor voltage v_c.
- Calculate the damping current by $i_p = v_c/R_p$.
- Adjust the modulations index m_a dynamically based on the calculated i_p.

By selecting a proper value for the emulated damping resistance R_p, the LC resonances can be effectively suppressed without causing any additional power losses.

The detected capacitor voltage v_c is composed of the fundamental component and all the harmonic components, and so is the calculated damping current i_p. In practice, the active damping control needs to damp out the LC resonances at all frequencies except the fundamental since the active damping of the fundamental component of i_p will interfere with the control of the rectifier dc output current [11].

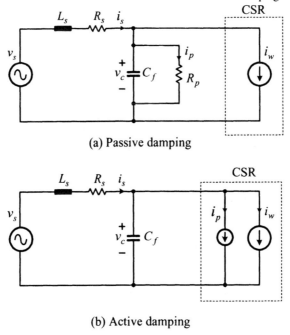

Figure 11.5-2 Realization of passive and active damping.

Figure 11.5-3 shows a block diagram of the active damping control algorithm. The capacitor voltage v_c is detected and then transformed to the synchronous reference frame. The transformation is performed by the abc/dq block based on a reference angle θ that is obtained from a digital phase-locked loop (PLL). The fundamental component of the capacitor voltage v_c in the stationary frame becomes a dc component in the synchronous frame. The dc component is then removed with a high-pass filter (HPF), and the remaining signal represents the capacitor harmonic voltages v'_c in the synchronous frame. This harmonic detection method is insensitive to the variations in the supply frequency, a desirable feature in practice. The high-pass filter is set to a low cutoff frequency, resulting in a minimal phase delay at the resonant frequencies.

The damping current reference i'_p is obtained by dividing v'_c by the desired damping resistance R_p. The calculated damping current i'_p is then converted to the damping modulation index m'_a by normalizing to the dc output current I_d. The damping modulation index m'_a is summed with the modulation index m''_a for dc current control in the space vector modulator for the active damping and dc current control.

It should be noted that space vector modulation schemes are recommended for the active damping control. The selective harmonic elimination (SHE) is an off-line scheme and may not be able to provide a fast adjustment of its modulation index. The SVM scheme discussed in Chapter 10 for the current source inverter can be directly applied to the CSR for active damping control.

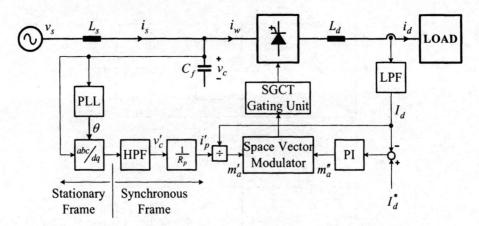

Figure 11.5-3 Block diagram of a vector-controlled CSR with active damping control.

11.5.4 LC Resonance Suppression

To investigate the performance of the active damping control illustrated in Fig. 11.5-3, computer simulations were carried out [14]. The PWM rectifier under investigation is powered by an ideal three-phase 4160-V utility supply. The parameters of the rectifier are given in Table 11.5-1. The emulated damping resistance R_p used in the controller is 1.5 pu.

Let's consider a case where the dc current reference I_d^* is reduced suddenly from 100% to 20% and the load resistance changes with the reference to maintain a constant load voltage of 50%. This simulates the dc load that the rectifier would see if it were used to feed an inverter-based drive system operating at a constant motor speed with a sudden decrease in load torque.

The simulated waveforms for the capacitor voltage v_c and line current i_s are shown in Fig. 11.5-4. The step change in the reference current forces the dc current control to react as quickly as possible, resulting in a fast transient on the ac side of

Table 11.5-1 Rectifier Ratings and Parameters Used in Simulation of a 1-MVA Rectifier

Rated power:	1 MVA
Rated voltage (line-to-line, rms):	4160 V
Rated line current (rms):	138 A
Total line inductance L_s:	0.17 pu
Filter capacitor C_f:	0.3 pu
Equivalent series resistance R_s:	0.002 pu
Active damping resistance R_p:	1.5 pu
LC resonant frequency ω_{res}:	4.43 pu (266 Hz)
Switching frequency f_{sw}:	540 Hz

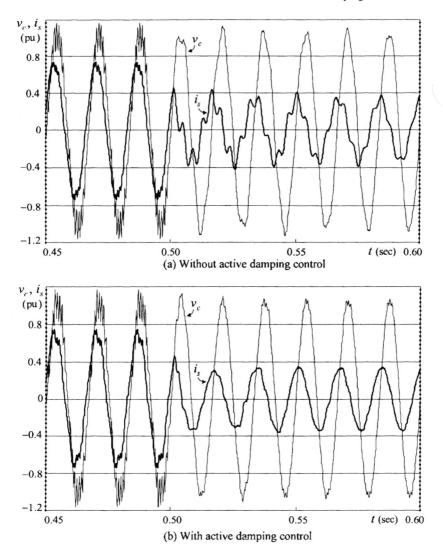

Figure 11.5-4 Simulated waveforms for 1MVA rectifier with a step reduction in I_d^*.

the rectifier. This transient may excite the natural resonance of the LC circuit. Without active damping control, the line current i_s and capacitor voltage v_c undergo oscillations during the dc side step transient as shown in part (a). Due to the light damping, the oscillations persist for many cycles of the fundamental frequency. However, when active damping is employed, the oscillations are suppressed within one cycle of the fundamental frequency as shown in part (b).

The active damping control is verified by experiments on a laboratory GTO current source rectifier with a digital controller. The parameters used in the experimental system are given in Table 11.5-2. The rectifier is controlled by a space vector modulation scheme with a switching frequency of 540 Hz. The supply voltage contains approximately 1% 5th harmonic voltage. Since the LC resonant frequency of 5.43 pu is close to the frequency of the 5th harmonic, the LC resonance can be excited by the 5th harmonic voltage in the supply and the 5th harmonic current generated by the rectifier.

Figure 11.5-5 shows the current and voltage waveforms measured from the laboratory rectifier without and with active damping control. The dc current reference is stepped from 75% to 20% of its rated value. The system response in part (a) exhibits oscillations due to the change in the dc current reference. In particular, the waveform of the line current i_s has two humps per half-cycle of the fundamental frequency and therefore contains a substantial amount of 5th harmonic component due to the LC resonance. With the active damping control implemented, the LC resonance is effectively suppressed as shown in part (b).

11.5.5 Harmonic Reduction

The active damping control can also reduce the line current distortion during the steady-state operation of the current source rectifiers [15]. To investigate the effect of LC resonances on line current THD, let's consider a 1-MW/4160-V rectifier with $L_s = 0.1$ pu, $C_f = 0.4$ pu, $R_s = 0.005$ pu, $L_d = 1.0$ pu, and $f_{sw} = 540$ Hz (SVM). The PWM current i_w of the rectifier normally contains a few percent of 5th and 7th harmonics produced by the space vector modulation. The natural frequency of the LC circuit is tuned exactly to the 5th harmonic ($\omega_{res} = 1/\sqrt{L_s C_f} = 5.0$ pu) on purpose, which is the worst case and should be avoided in practice. The simulated PWM current i_w, line current i_s and capacitor phase voltage v_c are shown in Fig. 11.5-6. Since there is no active damping present, the parallel resonant mode of the LC circuit is excited by the 5th harmonic current produced by the PWM rectifier. This is evident in the line current and capacitor voltage waveforms. The total harmonic distortion of the line current i_s is substantial (105%).

Figure 11.5-7 shows the line current THD profile for the rectifier system as a function of the series resistance R_s and resonant frequency f_{res}. The LC values are calculated using a 4:1 ratio of per unit filter capacitor C_f to total line inductance L_s. The dominant shape of this THD profile is a crest for the resonant mode of the LC

Table 11.5-2 Parameters in a Laboratory GTO Current Source Rectifier with Active Damping Control

Total line inductance L_s:	0.087 pu
Filter capacitor C_f:	0.39 pu
dc current inductance L_d:	0.87 pu
dc load resistance R_d:	1.2 pu
Active damping resistance R_p:	0.96 pu
LC resonant frequency ω_{res}:	5.43 pu (326 Hz)

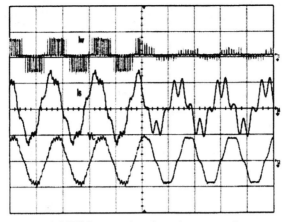

(a) Without active damping control

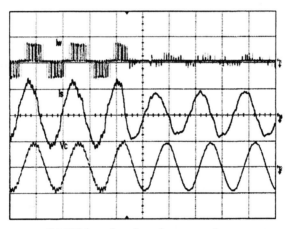

(b) With active damping control

i_w: 1.3 pu/div, i_s: 0.5 pu/div, v_c: 1.41 pu/div, Timebase: 10 ms/div.

Figure 11.5-5 Waveforms measured from a laboratory CSR with a step reduction in I_d^* at cursor ▲.

filter tuned to the fifth harmonic. The THD of the line current for a resonant frequency f_{res} of 5 pu and a series resistance of 0.5% is 105%. Considering an extreme case where the series resistance is 1.5% (impractically high), the line current THD is still 60%, which is too high to accept.

Figure 11.5-8 shows the effect of the active damping control on line current THD for the rectifier operating under the same conditions as in the previous case. When the LC filter is tuned to the 5th harmonic, an active damping resistance of 1.0 pu results in a line current THD of 5%, which is about 21 times lower than that without active damping control.

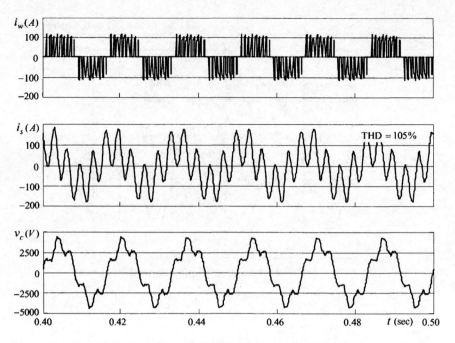

Figure 11.5-6 Simulated waveforms with the LC resonant mode tuned exactly to the 5th harmonic (worst case) without active damping control.

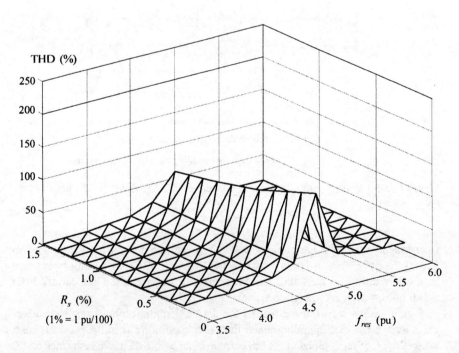

Figure 11.5-7 Line current THD versus series resistance R_s and resonant frequency f_{res} without active damping control.

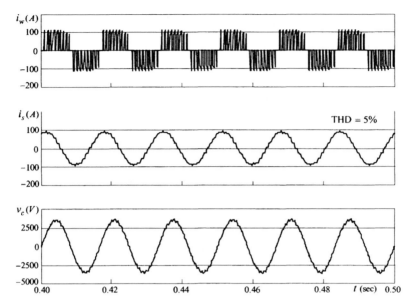

Figure 11.5-8 Simulated waveforms with the LC resonant mode tuned exactly to the 5th harmonic (worst case) with 1-pu active damping resistance.

The effect of active damping on the system sensitivity to the LC resonant mode is easily seen in the THD profile shown in Fig. 11.5-9. As the strength of active damping increases (i.e., a small value of R_p), the crest in the THD profile flattens. With an active damping resistance of 1 pu, the THD profile is almost a linear function of f_{res}. Active damping resistances below 1 pu are not shown in the figure because of the system instability caused by strong damping, where the significance of THD is lost due to nonperiodical waveforms.

11.5.6 Selection of Active Damping Resistance

As stated earlier, a small value of the active damping resistance R_p is preferred for minimizing the line current THD. However, if R_p is too small, the rectifier system may become unstable. The selection of the damping resistance is a complex issue since it is determined by many factors, including LC resonant modes, device switching frequency, dc current and voltage levels, magnitude of dominant harmonics, and value of modulation index. As a rule of thumb, the value of R_p for a high-power rectifier with a switching frequency of a few hundred hertz should not be lower than 1.0 pu.

To optimize the system performance, the damping resistance R_p should not be kept constant over the full operating range of the rectifier. Its minimum value can be determined by the stability requirements and may be used over most of the operating range. With the modulation index approaching to its maximum value, the damping resistance should increase accordingly [15].

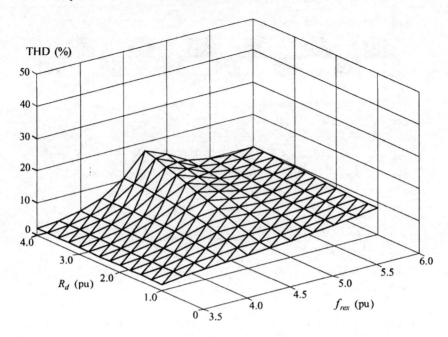

Figure 11.5-9 Line current THD versus damping resistance R_p and resonant frequency f_{res} ($R_s = 0.005$ pu).

11.6 SUMMARY

This chapter addresses four main issues for the current source rectifiers (CSR) used in high-power ac drives, including converter topologies, PWM schemes, power factor control, and LC resonance with active damping control. The single-bridge CSR topology features simple converter structure and low manufacturing cost while the dual-bridge CSR draws a sinusoidal current from the utility supply. Selective harmonic elimination (SHE) schemes for the two rectifiers are presented. To minimize the switching losses, these schemes are designed with a switching frequency of only 360 Hz or 420 Hz.

The current source rectifier requires a filter capacitor at its input terminals. To compensate the effect of the filter capacitor, a power factor control scheme with simultaneous delay angle and modulation index control is presented. This scheme can make the rectifier operate at the unity power factor or highest achievable power factor over the full operating range. Furthermore, the power factor control algorithm does not require any system parameters, which is desirable in practice.

The filter capacitor and line inductance constitute an LC resonant circuit. The LC resonances may be excited by voltage harmonics in the utility supply or current harmonics produced by the rectifier. An active damping control scheme is elaborated, which can effectively suppress the LC resonances during system transients and also reduce the line current distortion during the steady-state operation of the rectifier.

REFERENCES

1. N. Zargari, Y. Xiao, and B. Wu, A Near Unity Input Displacement Factor PWM Rectifier for Medium Voltage CSI Based AC Drives, *IEEE Industry Applications Magazine*, Vol. 5, No. 4, pp. 19–25, 1999.
2. E. P. Wiechmann, R. P. Burgos, and J. R. Rodriguez, Reduced Switching Frequency Active Front End Converter for Medium Voltage Current Source Drive Using Space Vector Modulation, *IEEE Symposium on Industrial Electronics (ISIE)*, pp. 288–293, 2000.
3. N. Zargari, S. C. Rizzo, et al., A New Current Source Converter Using a Symmetric Gate Commutated Thyrister (SGCT), *IEEE Transactions on Industry Application*, Vol. 37, No. 3, pp. 896–902, 2001.
4. B. Wu and F. DeWinter, Voltage Stress on Induction Motor in Medium Voltage (2300 V to 6900 V) PWM GTO CSI Drives, *IEEE Transactions on Power Electronics*, Vol. 12, No. 2, pp. 213–220, 1997.
5. Y. Xiao, B. Wu, F. DeWinter, et al., High Power GTO AC/DC Current Source Converter with Minimum Switching Frequency and Maximum Power Factor, *Canadian Conference on Electrical and Computer Engineering*, pp. 331–334, 1996.
6. J. R. Espinoza, G. Joos, J. I. Guzman, et al., Selective Harmonic Elimination and Current/Voltage Control in Current/Voltage Source Topologies: A Unified Approach, *IEEE Transactions on Industry Electronics*, Vol. 48, No. 1, pp. 71–81, 2001.
7. Y. Xiao, B. Wu, F. DeWinter, et al., A Dual GTO Current Source Converter Topology with Sinusoidal Inputs for High Power Applications, *IEEE Transactions on Industry Applications*, Vol. 34, No. 4, pp. 878–884, 1998.
8. F. DeWinter, N. Zargari, B. Wu, et al., Harmonic Eliminating PWM Converter, US Patent, #5,835,364, November 1998.
9. Y. Xiao, B. Wu, S. Rizzo, et al., A Novel Power Factor Control Scheme for High Power GTO Current Source Converter, *IEEE Transactions on Industry Applications*, Vol. 34, No. 6, pp. 1278–1283, 1998.
10. J. R. Espinoza and G. Joos, State Variable Decoupling and Power Flow Control in PWM Current-Source Rectifiers, *IEEE Transactions on Industrial Electronics*, Vol. 45, pp. 80–87, 1998.
11. J. Wiseman, B. Wu, and G. S. P. Castle, A PWM Current Source Rectifier with Active Damping for High Power Medium Voltage Applications, *IEEE Power Electronics Specialist Conference (PESC)*, pp. 1930–1934, 2002.
12. Y. Sato and T. Kataoka, A Current Type PWM Rectifier with Active Damping Function, *IEEE Industrial Applications Society Annual Meeting*, Vol. 3, pp. 2333–2340, 1995.
13. M. Salo and H. Tuusa, A Vector Controlled Current-Source PWM Rectifier with a Novel Current Damping Method, *IEEE Transactions on Power Electronics*, Vol. 15, pp. 464–470, 2000.
14. J. Wiseman, A Current-Source PWM Rectifier with Active Damping and Power Factor Compensation, Thesis, Master's in Engineering Science, University of Western Ontario, May 2001.
15. J. Wisewman and B. Wu, Active Damping Control of a High Power PWM Current Source Rectifier for Line Current THD Reduction, *IEEE Power Electronics Specialist Conference*, pp. 552–557, 2004.

APPENDIX: GATING ANGLES FOR CURRENT SOURCE RECTIFIERS

Table 1 Gating Angles for the Single-Bridge CSR with 5th and 7th Harmonic Elimination

m_a	0.1	0.2	0.3	0.4	0.5	0.6	0.7	0.8	0.9	1.0	1.03
					Degrees						
θ_1	-13.5	-11.9	-10.3	-8.60	-6.86	-5.00	-3.98	-0.67	2.17	6.24	7.93
θ_2	14.2	13.5	12.7	12.0	11.4	10.8	10.4	10.3	10.8	12.6	13.8
θ_3	43.6	42.2	40.9	39.5	38.0	36.6	35.1	33.6	32.1	30.5	30.0
θ_4	45.8	46.5	47.3	48.0	48.6	49.2	49.6	49.7	49.2	47.3	46.2
θ_5	73.5	71.9	70.3	68.6	66.9	65.0	63.0	60.7	57.8	53.8	52.1
θ_6	106.5	108.1	109.7	111.4	113.1	115.0	117.0	119.3	122.2	126.2	127.9
θ_7	134.2	133.5	132.7	132.0	131.4	130.8	130.4	130.3	130.8	132.6	133.8
θ_8	136.4	137.8	139.1	140.5	142.0	143.4	144.9	146.4	147.9	149.5	150.0
θ_9	165.8	166.5	167.3	168.0	168.6	169.2	169.6	169.7	169.2	167.3	166.2
θ_{10}	193.5	191.9	190.3	188.6	186.9	185.0	183.0	180.7	177.8	173.7	172.1
θ_{11}	256.4	257.8	259.1	260.5	262.0	263.4	264.9	266.4	267.9	269.5	270.0
θ_{12}	283.6	282.2	280.9	279.5	278.0	276.6	275.1	273.6	272.1	270.5	270.0

Table 2 Gating Angles for the Dual-Bridge CSR with 11th and 13th Harmonic Elimination

	Pattern A								Pattern B			
m_a	0.2	0.3	0.4	0.5	0.6	0.7	0.8	0.857	0.84	0.9	1.0	1.085
					Degrees							
θ_1	43.6	42.8	41.9	41.0	39.5	35.5	31.5	30.0	18.9	19.1	19.4	19.0
θ_2	46.2	46.6	46.8	46.4	44.4	38.7	35.1	34.2	18.6	19.6	21.1	21.7
θ_3	57.4	55.9	54.3	52.2	49.0	44.6	42.2	41.1	30.0	30.0	30.0	30.0
θ_4	122.6	124.1	125.7	127.8	131.0	135.4	137.8	138.9	33.9	34.8	36.4	38.3
θ_5	133.8	133.4	133.2	133.6	135.6	141.3	144.9	145.8	41.1	40.9	40.6	41.0
θ_6	136.4	137.2	138.1	139.0	140.6	144.5	148.5	150.0	138.9	139.1	139.4	139.0
θ_7	242.6	244.1	245.7	247.8	251.0	255.4	257.8	258.9	146.1	145.2	143.6	141.7
θ_8	253.8	253.4	253.2	253.6	255.6	261.3	264.9	265.8	150.0	150.0	150.0	150.0
θ_9	256.4	257.2	258.1	259.0	260.5	264.5	268.5	270.0	161.4	160.4	158.9	158.3
θ_{10}	283.6	282.8	281.9	281.0	279.5	275.5	271.5	270.0	161.1	160.9	160.6	161.0

Table 2 *Continued*

	Pattern A								Pattern B			
m_a	0.2	0.3	0.4	0.5	0.6	0.7	0.8	0.857	0.84	0.9	1.0	1.085
	Degrees											
θ_{11}	286.2	286.6	286.8	286.4	284.4	278.7	275.1	274.2	258.6	259.6	261.1	261.7
θ_{12}	297.4	295.9	294.3	292.2	289.0	284.6	282.2	281.1	266.1	265.2	263.6	261.7
θ_{13}	—	—	—	—	—	—	—	—	273.9	274.8	276.4	278.3
θ_{14}	—	—	—	—	—	—	—	—	281.4	280.4	278.9	278.3

Part Five

High-Power ac Drives

Chapter 12

Voltage Source Inverter-Fed Drives

12.1 INTRODUCTION

The voltage source inverter-fed medium-voltage (MV) drives have found wide application in industry. These drives come with a number of different configurations, each of which has some unique features. This chapter focuses on a few major voltage-source-based MV drives marketed by the world leading drive manufacturers. The advantages and limitations of these drives are analyzed.

12.2 TWO-LEVEL VSI-BASED MV DRIVES

It is well known that the two-level voltage source inverter is a dominant converter topology for low-voltage ($\leq$600 V) drives. This technology has now been extended to the MV drives, which are commercially available for the power rating up to a few megawatts [1].

12.2.1 Power Converter Building Block

Figure 12.2-1a shows a typical two-level inverter topology for the MV drive. There are three switch modules in series per inverter branch. Each switch module is composed of an IGBT device, gate driver, snubber circuit, parallel resistor R_p, and bypass switch as shown in Fig. 12.2-1b. This type of switch module is also known as **power converter building block** (PCBB) [2].

The gate driver receives a gate signal from the digital controller of the drive and generates conditioned gating pulses for the IGBT. It also detects the operating status of the IGBT and sends it back to the controller for fault diagnosis. The gate driver normally communicates with the drive controller through fiber-optic cables for electrical isolation and high noise immunity. A number of protection functions can be implemented in the gate driver as well, such as IGBT overvoltage and short-cir-

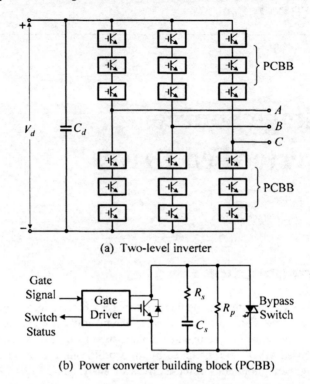

Figure 12.2-1 Two-level medium voltage inverter with PCBBs.

cuit protections. The active overvoltage clamping scheme presented in Chapter 2 for series-connected IGBTs can also be integrated into the gate driver.

Each IGBT switch is protected by an RC snubber network (C_s and R_s) from overvoltages at turn-off. The snubber also facilitates dynamic voltage equalization for the series connected devices during switching transients. Alternatively, the active overvoltage clamping scheme can be implemented instead of the snubber circuit. However, the use of active overvoltage clamping causes additional switching losses for the IGBTs as discussed in Chapter 2.

The snubber circuit provides an effective means of transferring the switching losses from the IGBT to the snubber resistor, leading to a lower junction temperature rise and better thermal management for the IGBTs. The snubber circuit also helps to reduce the dv/dt during the IGBT turn-off transients. The parallel resistor R_p shown in Fig. 12.2-1b is for static voltage sharing, and the function of the bypass switch will be discussed later.

12.2.2 Two-Level VSI Drive with Passive Front End

Figure 12.2-2 illustrates a typical configuration for the two-level VSI drive. A 12-pulse diode rectifier is employed as a front end for the reduction of line current har-

12.2 Two-Level VSI-Based MB Drives

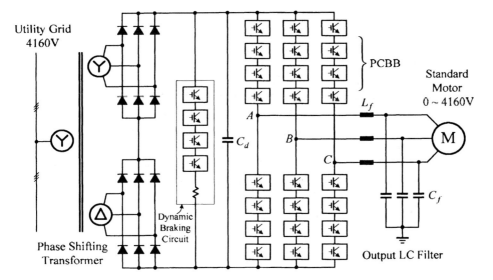

Figure 12.2-2 Typical two-level VSI drive with a passive front end.

monic distortion. For applications with more stringent harmonic requirements, the 12-pulse rectifier can be replaced by an 18- or 24-pulse diode rectifier. The detailed analysis on the multipulse diode rectifiers is given in Chapter 3.

The inverter is composed of 24 switch modules with four modules per inverter branch. Using 3300-V IGBTs, the two-level inverter is suitable for 4160-V (line-to-line) ac motors. The dynamic braking circuit in the dc link is optional. The dc capacitors are normally of oil-filled type instead of electrolytic type commonly used in the low-voltage VSI drives since the latter has limited voltage ratings (a few hundred volts each). The two-level inverter usually requires an LC filter at its output. The inverter can be controlled by either carrier-based modulation or space vector modulation scheme presented in Chapter 6.

The two-level voltage source inverter has the following features:

- **Modular structure using power converter building blocks (PCBBs).** The IGBT device, gate driver, bypass switch, and snubber circuits are integrated into a single switch module for easy assembly and mass production, leading to a reduction in manufacturing cost. The modular design also facilitates fast replacement of failed modules when the drive operates in the field.
- **Simple PWM scheme.** The conventional carrier-based sinusoidal modulation or space vector modulation scheme can be implemented for the inverter. Only six gate signals are required for the six groups of synchronous switches. The number of the gate signals does not vary with the number of the switches in series.
- **Active overvoltage clamping for series connected IGBTs.** The maximum dynamic voltage on the IGBT device at turn-off can be effectively clamped

by the gate driver. The IGBT can be safely protected from overvoltages caused by switching transients.

- **$N + 1$ provision for high reliability.** In applications where high system reliability is required, a redundant switching device ($N + 1$) can be added to each of the six inverter branches. When a switch module malfunctions during operation, the defective module can be shorted out by the bypass switch, and the drive is able to operate continuously at full load with a failed module.
- **Ease of dc capacitor precharging.** The dc capacitor in the two-level inverter needs only one pre-charging circuit. This is in contrast to the multilevel inverters where a multiple sets of precharging circuits are normally required.
- **Provision for four-quadrant operation and regenerative braking.** The multipulse diode rectifier can be replaced by an active front end with the same configuration as the inverter for four-quadrant operation or regenerative braking.

However, there are some drawbacks associated with the two-level voltage source inverter, including the following:

- **High dv/dt in the inverter output voltage.** Fast switching speed of IGBTs results in high dv/dt at the rising and falling edges of the inverter output voltage waveform. The dv/dt is particularly high for the two-level inverter employing series connected IGBTs switching in a synchronous manner. Depending on the magnitude of the dc bus voltage and switching speed of the IGBT, the dv/dt can well exceed 10,000 V/μs [3], which causes a number of problems such as premature failure of motor winding insulation, early bearing failure and wave reflections. More detailed explanation is given in Chapter 1.
- **Motor harmonic losses.** The two-level inverter usually operates at low switching frequencies, typically around 500 Hz, resulting in high harmonic distortion in the stator voltage and current. The harmonics produce additional power losses in the motor.
- **Common-mode voltages.** As discussed in Chapter 1, the rectification and inversion process in any converters generates **common-mode voltages** [4]. If not mitigated, these voltages would appear on the motor, causing premature failure of its winding insulation.

The first two problems can be effectively solved by adding a properly designed LC filter between the inverter output and the motor as shown in Fig. 12.2-2. With the use of the filter, the high dv/dt in the inverter output voltage is now applied to the filter inductor L_f instead of the motor. The insulation of the inductor should be properly designed for the high dv/dt. The LC filter is normally installed inside the drive cabinet and connected to the inverter with short cables to avoid wave reflections. The motor voltage and current can be made nearly sinusoidal by the filter, leading to low harmonic losses in the motor.

However, the use of the LC filter causes some practical consequences, including an increase in manufacturing cost, fundamental voltage drops, and circulating current between the filter and dc circuit. It may also cause LC resonances that can be excited by the harmonics in the inverter PWM voltages. The problem can be mitigated at the design stage by placing the LC resonance frequency below the lowest harmonic frequency [3]. The principle of the active damping control presented in the previous chapter can also be used for the suppression of the LC resonances.

The third problem can be effectively mitigated by the phase shifting transformer in Fig. 12.2-2, through which the common-mode voltages can be blocked. To ensure that the motor is not subject to any common-mode voltages, the neutral of the filter capacitor C_f is grounded directly or through an RC grounding network. In a three-phase balanced system, the neutral points of the capacitor and stator winding should have the same potential. Grounding one makes the other equivalently grounded.

It is worth mentioning that the use of the phase shifting transformer does not lead to the elimination of the common-mode voltage. Grounding the capacitor neutral essentially makes the common-mode voltage be transferred from the motor to the transformer [4]. The insulation system of the transformer should, therefore, be properly designed. The MV drive system is suitable for retrofit applications, where standard ac motors (which are not designed to withstand the common-mode voltages) are usually used.

12.3 NEUTRAL-POINT CLAMPED (NPC) INVERTER-FED DRIVES

The MV drive using three-level NPC inverter technology is marketed by a number of leading drive manufacturers [5–8]. Some manufacturers use GCTs in their drives while the others prefer IGBTs.

12.3.1 GCT-Based NPC Inverter Drives

Figure 12.3-1a shows a typical configuration of a three-level NPC inverter fed drive. A 12-pulse diode rectifier is adopted as a front end. The inverter consists of 12 reverse-conducting GCT devices and six clamping diodes. The mechanical assembly for one of the three inverter legs is illustrated in Fig. 12.3-1b, where four GCTs, two diodes, and a number of heatsinks can be assembled with just two bolts, leading to high power density and low package costs.

There are two di/dt clamp circuits, each composed of L_s, D_s, R_s, and C_s. One clamp circuit is in the positive dc bus for the switches in the upper half-bridge, and the other is in the negative dc bus for those in the lower half-bridge. With a few micro Henries for the di/dt limiting choke L_s, the rate of current rise during GCT turn-on transients can be limited to a certain value, typically below 1000 A/μs.

The neutral point Z of the NPC inverter can be connected to the midpoint X of the diode rectifier. Such a connection makes the total dc voltage equally divided between the two dc capacitors. The inverter neutral point voltage control in this case is no longer an issue.

Similar to the two-level VSI drive, an LC filter is usually installed at the inverter

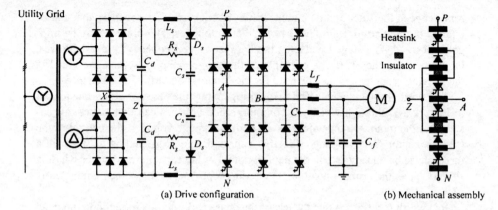

(a) Drive configuration   (b) Mechanical assembly

Figure 12.3-1 Typical configuration for a three-level NPC inverter-fed drive.

output terminals for sinusoidal outputs. The filter can also solve the dv/dt problems caused by fast switching of the GCT devices.

Figure 12.3-2 illustrates a three-level NPC drive with two more protection schemes added to the drive system of Fig. 12.3-1 [5, 6]. The drive is equipped with protection GCT switches S_d in the dc circuit for fuseless short-circuit protection. The di/dt limiting choke L_s limits the rate of rise of the dc current and facilitates a safe shutdown of the drive during a short-circuit fault.

As mentioned earlier, the common-mode voltages produced by the rectifier and inverter are transferred from the motor to the transformer for motor protection by grounding the neutral point of the filter capacitor C_f. To minimize the effect of the common-mode voltages on the transformer and its cables, a special common-mode choke L_{cm} is added to the dc link for the reduction of peak currents that cause charging and discharging of the capacitance of the cables that connect the transformer

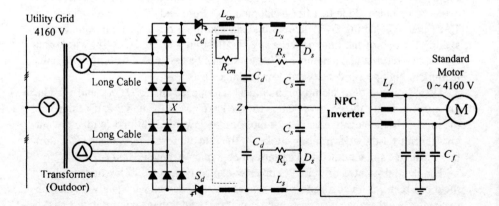

Figure 12.3-2 Three-level NPC drive with a common-mode choke for long transformer cables.

secondary windings to the rectifier. The choke has an auxiliary coil, to which a resistor R_{cm} is connected to suppress transient oscillations. With a properly designed L_{cm} and R_{cm}, the cable length can reach 300 m [6]. The transformer can then be placed outside the control room, which reduces the floor space and room cooling requirements as well.

It should be mentioned that the neutral point Z of the NPC inverter and midpoint X of the 12-pulse rectifier should be left unconnected due to the use of S_d and L_{cm}. As a result, the inverter neutral-point voltage should be tightly controlled as discussed in Chapter 8.

Table 12.3-1 gives the main specifications for the three-level NPC inverter fed

Table 12.3-1 Main Specifications for the Three-Level GCT-Based Drives

Drive System Specifications	Nominal input voltage	2300 V, 3300 V, 4160 V
	Output power rating	400–6700 HP (0.3–5 MW)
	Output voltage rating	0–2300 V, 0–3300 V, 0–4160 V
	Output frequency	0–66 Hz (up to 200 Hz optional)
	Drive system efficiency	Typically > 98.0% (including output filter losses but excluding transformer losses)
	Input power factor	> 0.95 (displacement power factor > 0.97)
	Output waveform	Sinusoidal (with output filter)
	Motor type	Induction or synchronous
	Overload capability	Standard: 10% for one minute every 10 minutes
		Optional: 150% for one minute every 10 minutes
	Cooling	Forced air or liquid
	Mean time between failure (MTBF)	> 6 years
	Regenerative braking capability	No
Control Specifications	Control scheme	Direct torque control (DTC)
	Dynamic speed error	< 0.4% without encoder
		< 0.1% with encoder
	Steady-state speed error	< 0.5% without encoder
		< 0.01% with encoder
	Torque response time	< 10 ms
Power Converter Specifications	Rectifier type	Standard: 12-pulse diode rectifier
		Optional: 24-pulse diode rectifier
	Inverter type	PWM, three-level NPC inverter
	GCT switching frequency	500 Hz
	Number of GCTs per phase	4
	Number of clamping diodes per phase	2
	Modulation technique	Hysteresis modulation generated by DTC scheme
	Inverter/rectifier switch failure mode	Non-rupture, non-arc

MV drive [5]. Without any GCTs connected in series, the drive is capable of powering ac motors with a rated voltage of 2300 V, 3300 V, or 4160 V. For the 4160-V drives, the GCTs rated at 5500 V can be selected. The rated power of the drive is typically in the range of 0.3 MW to 5 MW. It can be extended to 10-MW for 6600-V applications, where each switch position in the NPC inverter is replaced by two series-connected GCTs [7].

The switching frequency of the GCTs is typically around 500 Hz. However, the motor sees an equivalent switching frequency of 1000 Hz due to asynchronous switchings of the GCT devices, leading to a reduction in harmonic distortion and output filter size. This is one of the main features of the three-level NPC inverter-fed drive. In addition, the drive can operate at the medium voltages up to 4160 V without GCTs connected in series, which reduces the cost and increases the reliability of the drive due to the low component count.

12.3.2 IGBT-Based NPC Inverter Drives

The configuration of an IGBT-based three-level NPC drive is shown Fig. 12.3-3. The drive topology is essentially the same as that given in Fig. 12.3-2 except that neither the *di/dt* clamp circuits nor the protection switches are required. This is due to the fact that the rate of rise of IGBT anode current can be effectively controlled and the short-circuit protection can be fully implemented by the IGBT gate driver.

The main specifications of the three-level IGBT drive are given in Table 12.3-2. Depending on applications and customer requirements, the front end can be either the 12- or 24-pulse diode rectifier. A three-level IGBT-based NPC rectifier can also be used for the drives requiring four-quadrant operation or regenerative braking.

The MV drive can operate at the nominal utility/motor voltages of 2300 V, 3300

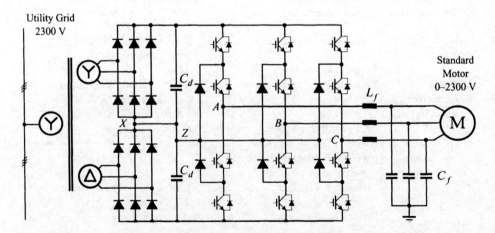

Figure 12.3-3 Three-level IGBT inverter-fed drive.

12.4 Multilevel Cascaded H-bridge (CHB) Inverter-Fed Drives

Table 12.3-2 Main Specifications for the Three-Level IGBT-Based Drives

Rectifier:	Standard: 12-pulse diode rectifier
	Optional: 24-pulse diode rectifier or active front end (PWM IGBT rectifier)
Displacement power factor ($\cos \varphi$):	> 0.96 (12-pulse diode rectifier)
Nominal utility/motor voltage:	2300 V, 3300 V, 4160 V, 6600 V
Output power rating:	0.8–2.4 MW @2300 V
	1.0–3.1 MW @3300 V
	1.3–4.0 MW @4160 V
	4.7–7.2 MW @4160 V (parallel converter configuration)
	0.6–2.0 MW @6600 V
Output voltage range:	0–2300 V, 0–3300 V, 0–4160 V, 0–6600 V
Output frequency:	0–100 Hz (standard)
Motor speed range:	1:1000 (with encoder)
Drive system efficiency:	Typically > 98.5%
	(at rated operating point, excluding transformer losses)

V, 4160 V and 6600 V. For the 2300-V applications, the NPC inverter is composed of 12 pieces of 3300-V IGBTs without devices in series. For the drives operating at higher voltages, two series-connected IGBTs can be used in each switch position in the inverter.

As listed in the table, the maximum power ratings of the drive are 2.4 MW at 2300 V, 3.1 MW at 3300 V and 4 MW at 4160 V. The power rating can be further increased to 7.2 MW at 4160 V with the two inverters operating in parallel. The operating voltage of the drive can be extended to 6600 V by adding a step-up autotransformer to the output of the 2300-V drive [9]. The leakage inductance of the autotransformer can also serve as the filter inductance, a viable solution for cost reduction.

12.4 MULTILEVEL CASCADED H-BRIDGE (CHB) INVERTER-FED DRIVES

The multilevel cascaded H-bridge (CHB) inverter is one of the popular inverter topologies for the MV drive [10, 11]. Unlike other multilevel inverters where high-voltage IGBTs or GCTs are used, the CHB inverter normally employs low-voltage IGBTs as a switching device in H-bridge power cells. The power cells are then connected in cascade to achieve medium voltage operation.

12.4.1 CHB Inverter-Fed Drives for 2300-V/4160-V Motors

The CHB inverters can be configured with different voltage levels. A seven-level cascaded H-bridge inverter fed MV drive is illustrated in Fig. 12.4-1. The phase-shifting transformer is an indispensable device for the CHB inverter. It provides

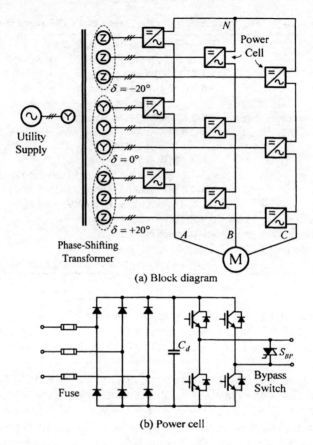

(a) Block diagram

(b) Power cell

Figure 12.4-1 Seven-level CHB drive with an 18-pulse diode rectifier.

three main functions: (a) isolated power supplies for the power cells, (b) line current THD reduction, and (c) isolation between the utility and the converter for common-mode voltage mitigation.

The phase-shifting transformer has three groups of secondary windings. Each group has three identical windings. The phase shift between any two adjacent winding groups is 20° for the seven-level CHB drive. Since each of the secondary windings is connected to a three-phase diode rectifier, this configuration is essentially an 18-pulse separate type diode rectifier discussed in Chapter 3.

The power cell is composed of a three-phase diode rectifier, a dc capacitor and a single-phase H-bridge inverter as shown in Fig. 12.4-1b. Each power cell is protected by fuses at the input and a bidirectional bypass switch S_{BP} at the output. The nominal output voltage of each power cell is typically 480 V (rms fundamental voltage). This design leads to the use of low-voltage components such as 1400-V IGBTs that are mass produced with a cost advantage over high-voltage (>1700 V) IGBTs. It is worth noting that the power cells should be insulated from each other

12.4 Multilevel Cascaded H-bridge (CHB) Inverter-Fed Drives

and from ground at medium-voltage levels even though they use low-voltage components.

Three power cells are cascaded at their ac output to form one line-to-neutral voltage of the three-phase system output. The inverter can produce seven distinct line-to-neutral voltage levels. The phase-shifted multicarrier sinusoidal modulation scheme presented in Chapter 7 is normally used in the CHB inverter. To boost the output voltage of each H-bridge, the 3rd harmonic injection method introduced in Chapter 6 can be adopted.

Table 12.4-1 summarizes the configuration of the multilevel CHB inverter-fed drives for medium-voltage applications. The topologies of the rectifier and inverter usually vary with the drive system operating voltages. For instance, with the utility/motor voltage of 3300 V, a 24-pulse diode rectifier and a nine-level CHB inverter can be selected. To reduce switching losses, the IGBT switching frequency $f_{sw,dev}$ is typically around 600 Hz. However, the equivalent inverter switching frequency $f_{sw,inv}$ is much higher due to the multilevel structure. The power rating of the drive is in the range of 0.3 MW to 10 MW.

The multilevel CHB inverter drive has a number of unique features:

- **Modular construction for cost reduction and easy repair.** The low-voltage power cells can be mass-produced for the multilevel CHB inverters operating at various medium voltages. Defective power cells can be easily replaced, which minimizes the downtime of the production-line.
- **Nearly sinusoidal output waveforms.** The CHB inverter is able to produce ac voltage waveforms with small voltage steps. The inverter normally does not require any filters at its output. The motor is protected from high dv/dt stresses and has minimal harmonic power losses.
- **Bypass function for improved system availability.** The faulty power cells can be bypassed and the drive can resume operation at reduced capacity with remaining cells. Although the bypass of defective cells may cause three-phase unbalanced operation for the motor, it allows the process to continue.

Table 12.4-1 Configurations of CHB Inverter-Fed MV Drives Using Low-Voltage IGBTs

Nominal Utility/ Motor Voltage (Volts)	Mulltipulse Diode Rectifier			Multilevel CHB Inverter					
	Rectifier Pulses	Secondary Windings	Transformer Secondary Cables	Power Cells	IGBTs	Voltage Levels (v_{AN})	Cell Output (Volts)	$f_{sw,dev}$ (Hertz)	$f_{sw,inv}$ (Hertz)
2300	18	9	27	9	36	7	480V	600	3600
3300	24	12	36	12	48	9	480V	600	4800
4160	30	15	45	15	60	11	480V	600	6000

- **Provision for N + 1 redundancy.** The drive system reliability can be improved by adding a redundant power cell to each of the inverter phase legs. When a power cell fails, it can be bypassed without causing reduction in the inverter output capacity.
- **Nearly sinusoidal line current.** This is mainly due to the use the multipulse diode rectifier.

There are a number of drawbacks for the multilevel CHB drive, including

- **High cost of phase-shifting transformer.** The multi-winding transformer is the most expensive device in the CHB drive. Its secondary windings should be specially designed such that the symmetry of leakage inductances is preserved for harmonic current cancellation.
- **Large number of cables.** The multilevel CHB inverter drive normally requires 27–45 cables connecting the power cells to the transformer. It is, therefore, expensive to place the transformer away from the drive. With the transformer installed inside the drive cabinet, the footprint of the drive increases, and so does the room cooling requirements.
- **Large component count.** The CHB inverter drive uses many low-voltage components, which potentially reduces reliability of the system.

12.4.2 CHB Inverter Drives for 6.6-kV/11.8-kV Motors

The operating voltage of the multilevel CHB inverter can be extended to 6600 V. A practical design for such a drive system is shown in Fig. 12.4-2, where two identical units of the seven-level CHB inverters are connected in cascade. The IGBTs used in the H-bridge power cells are typically rated at 1700 V, and each power cell produces a nominal voltage of 640 V. The power rating of the drive is in the range of 0.6 MW to 6 MW without IGBTs in series or parallel [12].

To drive the motors rated at 11.8 kV, the output voltage of each power cell can be increased to 1370 V. With five cells in cascade for the 11-level CHB inverter, its line-to-line voltages can reach 11.8 kV. High-voltage IGBTs should be used in this drive.

12.5 NPC/H-BRIDGE INVERTER-FED DRIVES

Figure 12.5-1 shows the drive configuration using 5-level NPC/H-bridge inverter [13]. The phase shifting transformer has three identical groups of secondary windings. Each group of secondary windings feeds a 24-pulse diode rectifier. The phase shift between any two adjacent secondary windings in each group is 15°. The neutral point of the NPC/H-bridges is tied to the midpoint of the rectifiers to avoid inverter neutral voltage deviation. The inverter phase voltage v_{AN} is composed of five

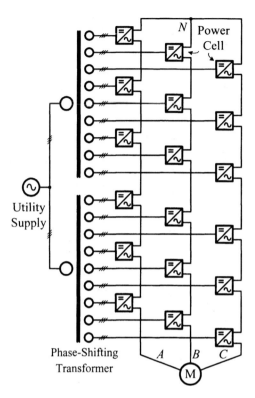

Figure 12.4-2 Topology of a 6600-V drive with two identical 7-level CHB inverters in cascade.

voltage levels while its line-to-line voltage v_{AB} has nine levels as discussed in Chapter 9.

The drive features very low ac line harmonic distortion, no switching devices in series, and low motor current THD. However, it requires a complex phase shifting transformer with 12 secondary windings. The drive also requires a *dv/dt* filter at the inverter output. The power rating of the drive using high-voltage IGBTs is in the range of 0.5 MW to 4.8 MW.

12.6 SUMMARY

This chapter presents various practical configurations of VSI-based MV drives, including two-level IGBT-fed drive, three-level NPC inverter drive, multilevel CHB inverter drive, and NPC/H-bridge inverter drive. The advantages and drawbacks of these drives are analyzed. Some practical problems are addressed, including high *dv/dt* stresses, wave reflections, common-mode voltages, and line/motor current distortion. Commonly used mitigation methods are also introduced.

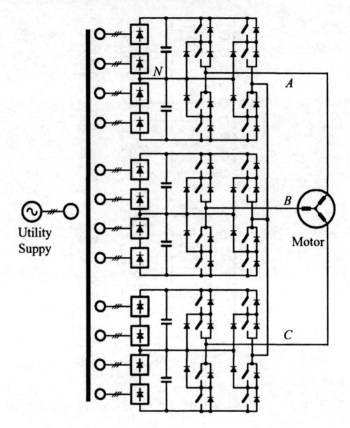

Figure 12.5-1 Five-level NPC/H-bridge inverter-fed MV drive.

REFERENCES

1. Y. Shakweh, New Bread of Medium Voltage Converters, *IEE Power Engineering Journal,* February Issue, pp. 12–20, 2000.
2. E. A. Lewis, Power Converter Building Blocks for Multi-megawatt PWM VSI Drives, *IEE Seminars on PWM Medium Voltage Drives,* pp. 4/1–4/19, 2000.
3. J. K. Steinke, Use of an LC Filter to Achieve a Motor-Friendly Performance of the PWM Voltage Source Inverter, *IEEE Transactions on Energy Conversion,* Vol. 14, No. 1, pp. 649–654, 1999.
4. S. Wei, N. Zargari, B. Wu, and S. Rizzo, Comparison and Mitigation of Common-Mode Voltage in Power Converter Topologies, *IEEE Industry Applications Society (IAS) Conference,* pp. 1852–1857, 2004.
5. S. Malik and D. Kluge, ACS1000 World's First Standard AC Drive for Medium-Voltage Applications, *ABB Review,* No. 2, pp. 4–11, 1998.
6. G. Brauer, A. Wirth, et al., Simulation Tools for the ACS1000 Standard AC Drive, *ABB Review,* No. 5, pp. 43–50, 1998.

7. J. P. Lyons, V. Blatkovic, et al., Innovation IGCT Main Drives, *IEEE Industry Applications Society Annual Meeting*, pp. 2655–2661, 1999.
8. R. Sommer, A. Mertens, et al., New Medium Voltage Drive Systems Using Three-Level Neutral Point Clamped Inverter with High Voltage IGBT, *IEEE Industry Applications Society Annual Meeting*, pp. 1513–1519, 1999.
9. R. Sommer, A. Mertens, et al., Medium Voltage Drive System with NPC Three-Level Inverter Using IGBTs, *IEE Seminars on PWM MV Drives*, pp. 3/1–3/5, 2000.
10. P. W. Hammond, A New Approach to Enhance Power Quality for Medium Voltage AC Drives, *IEEE Transactions on Industry Applications*, Vol. 33, No. 1, pp. 202–208, 1997.
11. R. H. Osman, A Medium Voltage Drive Utilizing Series-Cell Multilevel Topology for Outstanding Power Quality, *IEEE Industry Applications Society (IAS) Conference*, pp. 2662–2669, 1999.
12. *TOSVERT-MV AC Drive System*, Toshiba Product Brochure (DKSA-13043), 19 pages, 2002.
13. *Dura-Bilt5i MV—Medium Voltage AC Drive Topology Comparisons & Feature-Benefits*, GE Toshiba Automation Systems, April 8, 2003.

Chapter 13

Current Source Inverter-Fed Drives

13.1 INTRODUCTION

The current source inverter (CSI) technology is well-suited for medium-voltage (MV) drives. The CSI drive generally features simple converter structure, motor-friendly waveforms, inherent four-quadrant operation capability and reliable fuseless short-circuit protection. The main drawback lies in its limited dynamic performance. As indicated in Chapter 1, the majority (around 85%) of the installed MV drives are for high-power fans, pumps, and compressors where high dynamic performance is usually not a prime importance. Therefore, the CSI drive is perfectly fitted into this type of applications. The power rating of the PWM current source drives is normally in the range of 1–10 MW and can be further increased with parallel inverters. For a higher power rating up to 100 MW, the load-commutated inverter (LCI) is a preferred choice, with which the voltage source inverters normally cannot compete in terms of the cost and energy efficiency of the system.

This chapter presents a number of CSI drives with various front-end converters such as single-bridge PWM rectifier, dual-bridge PWM rectifier, and phase-shifted SCR rectifiers. The advantages and limitations of these drives are analyzed and the main specifications are provided. A new-generation transformerless CSI drive for standard ac motors is also presented. This CSI drive, reflecting the latest technology in the field, has gained wide acceptance over the past few years. The chapter ends with an introduction to LCI synchronous motor drives.

13.2 CSI DRIVES WITH PWM RECTIFIERS

13.2.1 CSI Drives with Single-Bridge PWM Rectifier

Figure 13.2-1 shows a typical CSI drive using a single-bridge PWM current source rectifier (CSR) as a front end. The rectifier and inverter have an identical topology using symmetrical GCTs. With the GCT voltage rating of 6000 V and two GCTs con-

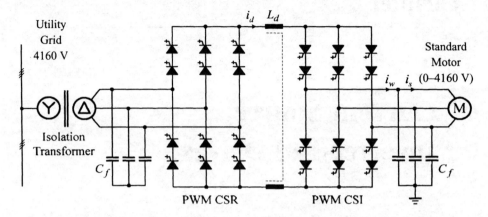

Figure 13.2-1 A typical 4160-V CSI drive with a PWM rectifier and inverter.

nected in series in each of the converter branches, the drive is capable of operating at the utility voltage of 4160 V (line-to-line). For higher operating voltages, the number of series connected devices can be increased while the converter topology remains unchanged. For example, three GCTs can be in series for the 6600-V drives.

The PWM rectifier can use either the space vector modulation (SVM) or selective harmonic elimination (SHE) schemes. The SHE scheme is preferred due to its superior harmonic performance. With a switching frequency of 420 Hz and the utility frequency of 60 Hz, a maximum of three low-order current harmonics (the 5th, 7th, and 11th) can be eliminated. The other high-order harmonics can be attenuated by the filter capacitor, leading to a very low line current THD.

The dc current i_d can be adjusted by the rectifier delay angle or modulation index control presented in Chapter 11. The delay angle control produces a lagging power factor that can compensate the leading current produced by the filter capacitor, resulting in an improved power factor. Furthermore, when the CSI drive is for fan or pump applications, its input power factor can be made near unity over a wide speed range [1]. This can be realized by using the SHE scheme with a delay angle control and properly selecting the line- and motor-side filter capacitors. This control scheme features simple design procedure, easy digital implementation and reduced switching loss (due to the lack of bypass operation), and therefore it is a practical control scheme for the PWM CSI drive.

A typical switching pattern arrangement for the inverter is illustrated in Fig. 13.2-2, where f_{sw} is the GCT switching frequency, f_1 is the inverter fundamental output frequency, and N_p is the number of pulses per half-cycle of the inverter PWM current i_w. The SHE scheme (without bypass pulses) is used at high inverter output frequencies while the trapezoidal pulse width modulation (TPWM) scheme is utilized when the motor operates at low speeds. As discussed in the previous chapter, the SHE scheme has a better harmonic performance than TPWM, but it is difficult, if not impossible, to eliminate more than four harmonics in i_w simultane-

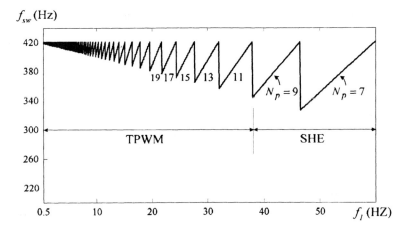

Figure 13.2-2 Switching frequency f_{sw} versus inverter output frequency f_1.

ously. The switching frequency of the inverter is normally below 500 Hz in industrial CSI drives.

It is worth noting that the GCT device is capable of operating at higher switching frequencies. The main reasons for limiting the switching frequency is the GCT's high thermal resistance that prevents efficient heat transfer from the device to its heatsink. In addition, the device switching frequency should be kept low for switching loss reduction. Prior to the advent of the GCT device, GTO was the main switching device for the high-power CSI drive, where the device switching frequency was around 200 Hz [2].

Similar to the VSI drives, the rectification and inversion process in the CSI drive generates common-mode voltages. If not mitigated, the common-mode voltage would appear on the motor, causing premature failure of its winding insulation. The problem can be effectively solved by introducing an isolation transformer and grounding the neutral of the inverter filter capacitor as shown in Fig. 13.2-1. In doing so, the motor is not subject to any common-mode voltages. Therefore, the CSI drive is suitable for retrofit applications, where standard ac motors are usually used.

The CSI MV drive has the following features:

- **Simple converter topology with low component count.** The converter topology is simple and independent of operating voltages. Both converters use symmetrical GCT devices that do not require antiparallel freewheeling diodes, leading to a minimum component count.
- **Motor friendly waveforms.** The motor voltage and current waveforms are close to sinusoidal and does not contain any voltage steps with high dv/dt. These waveforms are compatible with standard ac motors without derating.
- **High input power factor.** The CSI drive using the PWM rectifier as a front end has a minimum input power factor of 98% over a wide speed range [3].

This is in contrast with the conventional CSI drive with an SCR rectifier, where the power factor varies with the drive operating conditions.

- **Simple PWM scheme.** The SHE and TPWM schemes are normally used in the current source converters. These schemes are much simpler than those for multilevel voltage source inverters. Only six gate signals are required for the six groups of synchronous switches per converter. The number of the gate signals does not vary with the number of the switches in series.

- **Reliable fuseless short circuit protection.** In case of short circuit at the inverter outputs, the rate of rise of the dc current is limited by the dc choke, providing sufficient time for the drive controller to react. For a quick system shutdown, the rectifier can produce a negative dc voltage, forcing the dc current to fall quickly. The drive can be safely shut down at the moment when the dc current falls to zero. The drive does not need fuses for the short circuit or over-current protection.

- **N + 1 provision for high reliability.** In certain applications where high system reliability is required, a redundant switching device (N + 1) can be added to each of the six converter branches. Since the GCT device is normally short-circuited at failure, the CSI drive is able to operate continuously at its full capacity with a failed device.

- **Long input and output cables.** There are virtually no limit on the length of the cables connecting the transformer to the rectifier or connecting the inverter to the motor due to near sinusoidal input and output waveforms.

- **Four-quadrant operation and regenerative braking capability.** The power flow in the CSI drive is bidirectional. No additional components are required for four-quadrant operation and dynamic braking.

The main drawbacks of the CSI drive include the following:

- **Limited dynamic performance.** This is mainly due to the use of a dc choke that limits the rate of dc current changes. The dynamic performance of the drive can be significantly improved when the inverter output current is directly regulated by modulation index instead of dc current adjustment through the rectifier. However, this method is seldom used in practice because (a) it causes additional power losses due to the bypass operation and (b) the majority of the MV drives are for fans, pumps, and compressors, where high dynamic performance is usually not a prime importance.

- **Potential LC resonances.** The line- and motor-side filter capacitors constitute resonant modes with the line and motor inductances. The line-side LC resonances can be effectively avoided by properly sizing the filter capacitor and line reactor such that the resultant LC resonant frequency is lower than that of the lowest harmonic produced by the rectifier. This approach can be equally applied to suppress the motor-side LC resonances. In addition, the active damping control discussed in Chapter 11 also provides an effective means for LC resonance suppression.

Figure 13.2-3 Picture of a 4160-V drive with PWM GCT rectifier and inverter. Courtesy of Rockwell Automation Canada.

Figure 13.2-3 shows a picture of a 4160-V CSI drive system. The left cabinet contains an advanced digital controller for the drive system and the line- and motor-side filter capacitors. The middle cabinet houses two identical PWM converters, one for the rectifier and the other for the inverter. Each converter is composed of 12 pieces of 6000-V symmetrical GCTs. All the GCT devices in both converters are installed in six power-cage modules designed for quick device assembly and replacement. The dc choke and air-cooling system are installed in the right cabinet.

The main specifications for the MV CSI drives are given in Table 13.2-1 [3]. They are divided into three groups. The drive system specifications include typical voltage/power ratings, efficiency, input power factor, line current THD, and ride-through capability. The control specifications include the control scheme, parameter tuning method, speed regulation, and bandwidth of the controllers. The converter type, modulation techniques, and switching devices are listed as the converter specifications.

13.2.2 CSI Drives for Custom Motors

In certain applications where a new motor is to be ordered, the total cost of the drive can be reduced by replacing the isolation transformer with a three-phase line reactor L_s as shown in Fig. 13.2-4. The motor in this case must be custom-made with en-

Table 13.2-1 Main Specifications for the Medium-Voltage CSI Drives

Drive System Specifications	Nominal input voltage	2300 V, 3300 V, 4160 V, 6600 V
	Output power rating	200–9000 HP (150–6700 kW)
	Output voltage rating	0–2300 V, 0–3300 V, 0–4160 V, 0–6600 V
	Output frequency	0.2–85 Hz
	Drive system efficiency	> 96.0% (including transformer losses)
	Input power factor	Typically >0.98 (with PWM rectifier)
	Line current THD	Typically <5% (with PWM rectifier)
	Output waveform	Near sinusoidal current and voltage
	Motor type	Induction or synchronous
	Overload capability	150%—one minute (standard)
	Power loss ride-through	Five cycles
	Cooling	Forced air or liquid
	Regenerative braking capability	Inherent. No additional hardware or software required.
Control Specifications	Control scheme	Direct-field-oriented (vector) control
	Shaft encoder	Optional
	Parameter tuning method	Automatic self-tuning control
	Speed regulation	0.1% without shaft encoder
		0.01% with shaft encoder
	Speed regulator bandwidth	5–25 radians/second
	Torque regulator bandwidth	15–50 radians/second
Power Converter Specifications	Rectifier type	6-pulse SCR, 18-pulse SCR or PWM GCT
	Number of rectifier switches per phase	2@2400 V, 4@3300 V/4160 V, 6@6600 V
	Inverter type	PWM current source
	Number of inverter switches per phase	2@2400 V, 4@3300V/4160 V, 6@6600 V
	GCT/SCR peak off-state voltage rating	6000 V
	Modulation technique	Rectifier: SHE
		Inverter: SHE combined with TPWM
	GCT switching frequency	Rectifier: < 500 Hz
		Inverter: < 500 Hz
	Inverter/rectifier switch failure mode	Non-rupture, non-arc

13.2 CSI Drives with PWM Rectifiers

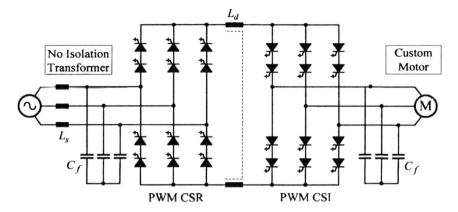

Figure 13.2-4 PWM CSI drive without isolation transformer.

hanced insulation to withstand common-mode voltage stresses. The additional benefits of such a drive include reduced footprint and operating cost due to the elimination of the isolation transformer.

13.2.3 CSI Drives with Dual-Bridge PWM Rectifier

Figure 13.2-5 illustrates a 4160-V CSI drive with a dual-bridge PWM rectifier. As discussed in Chapter 11, the rectifier uses SHE scheme to eliminate the 11th and 13th harmonics and phase shifting transformer to cancel the 5th, 7th, 17th, and 19th harmonics. The other remaining high-order harmonics can be absorbed by the filter

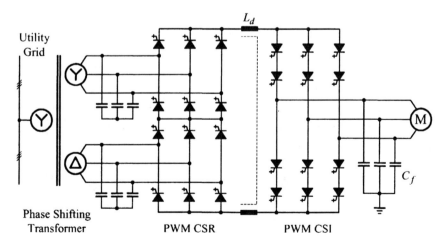

Figure 13.2-5 A 4160-V CSI drive with dual-bridge PWM rectifier for sinusoidal inputs.

capacitors. As a result, the line current waveform is virtually sinusoidal [4, 5]. This drive is suitable for applications that require extremely low line current distortion.

13.3 TRANSFORMERLESS CSI DRIVE FOR STANDARD ac MOTORS

As discussed earlier, one of the major problems in both CSI- and VSI-based MV drives is the common-mode voltage, which would cause premature failure of motor winding insulation if not mitigated. The problem can be solved by introducing an isolation or phase-shifting transformer, but this solution brings a few drawbacks such as high manufacturing cost (the transformer accounts for around 20% t0 25% of total drive cost), increased drive size and weight, and high operating cost due to transformer losses. Another solution is to use a custom-made ac motor with an increased insulation level. This solution, however, cannot be used for retrofit applications where standard motors are already in use. This section presents a new-generation transformerless CSI drive that can essentially solve all the problems mentioned above.

13.3.1 CSI Drive Configuration

Figure 13.3-1 shows the schematic of a transformerless CSI drive, where an integrated dc choke with a single magnetic core and four coils is used. The choke can provide two inductances, a differential inductance L_d that is inherently required by the CSI drive and a common-mode inductance L_{cm} that can block the common-mode voltage. The use of the integrated dc choke leads to the elimination of the isolation transformer, resulting in a significant reduction in manufacturing and operating costs. To ensure that the motor is not subject to any common-mode voltages, the

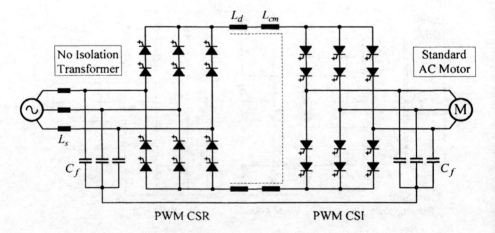

Figure 13.3-1 Transformerless CSI drive with an integrated dc choke.

neutral points of the line- and motor-side filter capacitors are connected together directly or through a small RL network.

13.3.2 Integrated dc Choke for Common-Mode Voltage Suppression

Figure 13.3-2 shows the structure of the integrated dc choke and its connection diagram [6]. The choke has three legs of equal cross-sectional area and two air gaps of equal length placed in the outer legs. The differential coils x_1 and x_2 are wound over the outer legs while the common-mode coils y_1 and y_2 are placed on the center leg. The number of turns of the differential and common-mode coils can be adjusted separately to produce required differential and common-mode inductances.

The connection of the integrated choke in the dc link of the drive is shown in Fig. 13.3-2b, where the differential coil x_1 is in series with the common-mode coil y_1 in the positive dc bus while coils x_2 and y_2 are in series in the negative bus. To avoid the motor winding subject to any common-mode voltages, the neutral point of the filter capacitor C_f is grounded. The dc current i_{dc} contains two components, a differential current i_d and a common-mode current i_{cm}. The differential current i_d flows through the positive dc bus and returns to the rectifier from the negative bus.

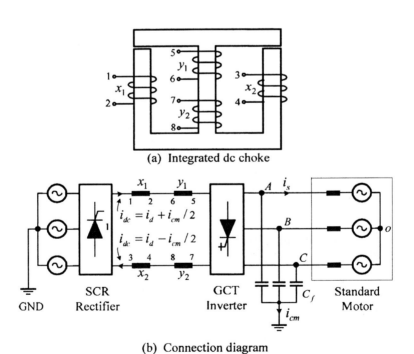

Figure 13.3-2 Structure of the integrated choke and its connection diagram in an experimental setup.

278 Chapter 13 Current Source Inverter-Fed Drives

The common-mode current i_{cm} is equally split in the dc circuit, one half in the positive bus and the other half in the negative dc bus due to the symmetrical arrangement of the choke and converters. The common-mode current i_{cm} flows through the rectifier, inverter, and filter capacitor neutral to ground, and its magnitude is limited by the common-mode inductance of the choke. With the same number of turns for all four coils, the ratio of L_{cm} to L_d can reach 2.25 [6].

The concept of the integrated choke is verified on a low-power laboratory CSI drive with a six-pulse SCR rectifier shown in Fig. 13.3-2. The inverter operates at a switching frequency of 180 Hz with the filter capacitor of $C_f = 0.31$ pu. The motor operates near rated speed under no load conditions, at which the common-mode voltage reaches its highest level. The fundamental frequency of the drive is 57.7 Hz at which the effect of the beat frequency (the frequency difference between the utility supply and the inverter) can be observed.

Figure 13.3-3 shows the measured waveforms of the drive with the integrated dc choke installed, where v_{AB} is the motor line-to-line voltage, i_s is the stator cur-

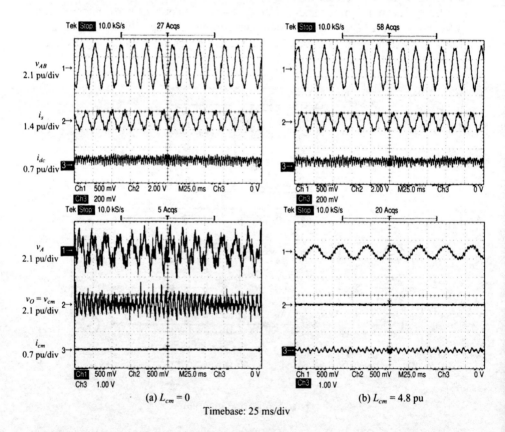

Figure 13.3-3 Waveforms measured from a laboratory CSI drive with an integrated dc choke.

rent, i_{dc} is the dc current, v_A is the motor line-to-ground voltage, v_O is the motor neutral-to-ground voltage, and i_{cm} is the common current. The scillogram in Fig. 13.3-3a shows the waveforms of the drive with $L_d = 1.0$ pu and $L_{cm} = 0$ (the common-mode coils y_1 and y_2 are not used in this testing). The neutral point of the filter capacitor C_f is left open. The neutral-to-ground voltage v_O, which should be zero when the motor is powered by a three-phase utility supply, is the total common-mode voltage v_{cm} generated by the rectifier and inverter. It contains triplen harmonics, which are of zero sequence components with the 3rd and 6th being dominant. The line-to-ground voltage v_A is composed of two components: the motor phase (line-to-neutral) voltage v_{AO} and its neutral-to-ground (common-mode) voltage v_O. Obviously, with the common-mode voltage superimposed on the phase voltage, v_A is severely distorted with high peaks, causing damages to the motor insulation system.

With the common-mode coils connected in the drive ($L_{cm} = 4.8$ pu) and the capacitor neutral grounded, the measured waveforms are shown in Fig. 13.3-3b. Comparing with Fig. 13.3-3a, we can make the following observations:

- The motor voltage v_{AB}, stator current i_s and dc current i_{dc} are essentially the same as those in Fig. 13.3-3a. It implies that the common-mode inductance provided by the integrated dc choke does not affect the operation of the drive.
- The waveform for v_A is close to sinusoidal, free of common-mode voltages. The common-mode voltage is fully blocked by the integrated dc choke. This observation can be verified by the waveform for v_O ($v_O = v_{cm}$), which is essentially zero.
- A small amount of common-mode current i_{cm} flows to ground through the filter capacitor. For the CSI drive with a PWM rectifier shown in Fig. 13.3-1, the common-mode current i_{cm} flows between the rectifier and inverter through the two capacitor neutrals, and therefore it will not cause any neutral current in the utility supply.

Finally, it should be noted that the design of the integrated dc choke is not unique. The above example is given to illustrate the concept of using an integrated dc choke in CSI drives for the motor common-mode voltage mitigation. Better designs are currently being implemented in the medium-voltage CSI drives, but they cannot be released here. Another valid design is given in reference 7.

13.4 CSI DRIVE WITH MULTIPULSE SCR RECTIFIER

13.4.1 CSI Drive with 18-Pulse SCR Rectifier

The multipulse SCR rectifiers presented in Chapter 4 can also be used as a front end for the CSI drive. Among the various rectifier topologies, the 18-pulse SCR rectifier is a preferred choice due to its high performance-to-price ratio. For a given voltage and power rating, the 12-pulse rectifier has a cost advantage over the 18-pulse,

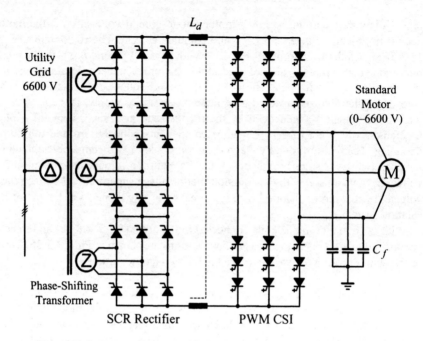

Figure 13.4-1 A typical 6600-V CSI drive with an 18-pulse SCR rectifier.

but its line current THD may not meet the harmonic guidelines set by most standards. The 24-pulse rectifier has a better line current THD than the 18-pulse, but it costs more due to the increased number of SCR devices and complex phase shifting transformer.

Figure 13.4-1 shows a typical CSI-fed drive with an 18-pulse SCR rectifier. The rectifier has three units of 6-pulse rectifiers in cascade while the inverter uses three series-connected GCT devices per branch. With the voltage rating of 6000 V for SCRs and GCTs, the drive is capable of operating at the utility voltage of 6600 V (line-to-line). The CSI drive can meet the harmonic requirements set by IEEE standard 519-1992, and it can also be used in retrofit applications with standard ac motors.

13.4.2 Low-Cost CSI Drive with 6-Pulse SCR Rectifier

Figure 13.4-2 shows a low-cost CSI drive with a 6-pulse SCR rectifier. The drive provides the most economical solution in certain applications. An optional harmonic filter can be used to reduce the line current THD. The installation cost can be further reduced for new motor applications where the isolation transformer can be replaced by a three-phase line reactor.

The low-cost CSI drive can also serve as a soft starter for the motors requiring a high starting torque. The conventional SCR based variable-voltage motor starters

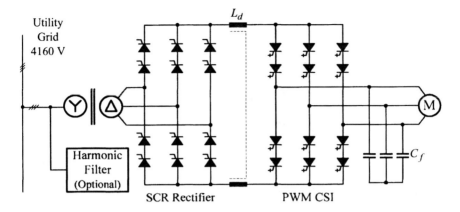

Figure 13.4-2 A low-cost CSI drive with 6-pulse SCR rectifier.

are incapable of producing a high starting torque due to the reduced starting voltage and the fixed supply frequency.

13.5 LCI DRIVES FOR SYNCHRONOUS MOTORS

The load-commuted inverter (LCI) fed synchronous motor drives are often found in high-power applications with a power rating up to 100 MW. The LCI drive uses low-cost SCR thyristors in the rectifier and inverter, leading to a substantial cost reduction.

13.5.1 LCI Drives with 12-Pulse Input and 6-Pulse Output

Figure 13.5-1 illustrates the typical configuration of an LCI fed synchronous motor drive. It uses a 12-pulse SCR rectifier with a dc link choke to provide a controllable dc current to the inverter. The number of series connected SCRs depends on the voltage ratings of the SCRs and the utility supply [8].

It is well known that the line current of the 12-pulse SCR rectifier does not contain the 5th or 7th harmonics, but it does have a large amount of the 11th and 13th harmonics. Therefore, a harmonic filter is often required for the LCI drive. The filter is of an LC series-resonant type, normally tuned at the 11th and 13th harmonics. By a proper selection of LC parameters, the turned filter can also serve as a power factor compensator (PFC).

The main features of the LCI-fed synchronous drive include low cost, high efficiency, reliable operation, and inherent regenerative braking capability. The main drawbacks are high torque pulsation, slow transient response, and variable input power factor.

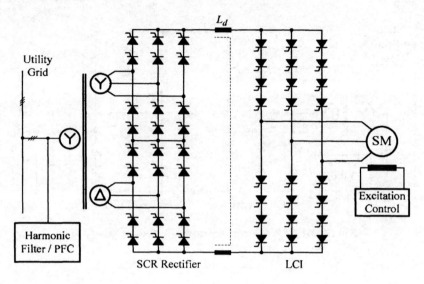

Figure 13.5-1 An LCI drive system with 12-pulse input and 6-pulse output.

13.5.2 LCI Drives with 12-Pulse Input and 12-Pulse Output

The harmonic currents in the LCI drive with 6-pulse output cause harmonic torques and power losses in the synchronous motor. The harmonic torque may excite mechanical vibrations due to the mechanical resonant modes of the drive system. These problems can be mitigated by the LCI drive with a 12-pulse output as shown in Fig. 13.5-2.

The synchronous motor has two sets of separate stator windings with a 30° electrical phase shift between them. The dominant harmonic torque produced by the 5th and 7th harmonic currents in the two stator windings can be canceled by the six-phase motor. As a result, the torque pulsation is substantially reduced and the mechanical stress on the shaft train is also decreased.

13.6 SUMMARY

This chapter focuses on a number of CSI drives with various front-end converters including single-bridge PWM rectifier, dual-bridge PWM rectifier, and phase-shifted SCR rectifiers. These medium-voltage drives are widely used in industry for high-power fans, pumps, and compressors. The advantages and limitations of the drives are analyzed and the main specifications are provided.

A new-generation transformerless PWM CSI drive for standard ac motors is also presented. The drive uses an integrated dc choke instead of isolation or phase-shift-

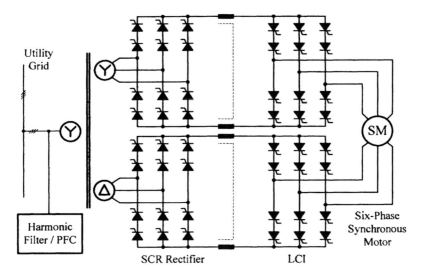

Figure 13.5-2 LCI drive with 12-pulse input and 12-pulse output.

ing transformer to block the common-mode voltages, leading to a substantial reduction in manufacturing cost. This CSI drive reflects the latest technology in the field and has gained wide acceptance over the past few years.

REFERENCES

1. Y. Xiao, B. Wu, and N. Zargari, Design of Line/Motor Side Capacitors for PWM CSR-CSI Drives to Achieve Optimal Power Factor in High Power Fan/Pump Applications, *IEEE Applied Power Electronics Conference*, pp. 333–337, 1997.
2. B. Wu, S. Dewan, and G. Slemon, PWM-CSI Inverter Induction Motor Drives, *IEEE Transactions on Industry Applications,* Vol. 28, No. 1, pp. 64–71, 1992.
3. PowerFlex 7000 MV Drive Brochure, Rockwell Automation Canada Inc., 2002.
4. Y. Xiao, B. Wu, F. DeWinter, et al., A Dual GTO Current Source Converter Topology with Sinusoidal Inputs for High Power Applications, *IEEE Transactions on Industry Applications,* Vol. 34, No. 4, pp. 878–884, 1998.
5. F. DeWinter, N. Zargari, B. Wu, et al., Harmonic Eliminating PWM Converter, US Patent 5,835,364, November, 1998.
6. B. Wu, S. Rizzo, N. Zargari, et al., Integrated dc Link Choke and Method for Suppressing Common-Mode Voltage in a Motor Drive, US Patent 6,617,814 B1, September 9, 2003
7. P. Hammond, Apparatus and Method to Reduce Common Mode Voltage from Current Source Drive, US Patent 5,905,642, November 1997.
8. R. Bhatia, H. U. Krattiger, A. Bonanini, et al., Adjustable Speed Drive with a Single 100-MW Synchronous Motor, *ABB Review,* No. 6, pp. 14–20, 1998.

Chapter 14

Advanced Drive Control Schemes

14.1 INTRODUCTION

Two advanced control schemes, field-oriented control (FOC) and direct torque control (DTC), have become the industrial standards in high-power medium-voltage (MV) drives. This is in view of the following: (a) The control schemes offer superior dynamic performance; (b) their algorithms can be efficiently implemented in real time by digital processors; and (c) the cost difference between the digital implementations of advanced and low-performance control schemes is minimal.

This chapter focuses on FOC and DTC schemes for induction motor drives. It starts with an introduction to reference frame transformation, followed by dynamic models of the induction motor. Various field-oriented control schemes for the voltage and current source drives are presented. The emphasis is on the rotor flux oriented control scheme due to its simplicity and wide acceptance in the MV drives. The direct torque control is also elaborated. Important concepts are illustrated with computer simulations and experiments. The chapter ends with a comparison between the FOC and DTC schemes.

14.2 REFERENCE FRAME TRANSFORMATION

The use of reference frame theory can simplify the analysis of electric machines and also provide a powerful tool for the digital implementation of sophisticated control schemes for ac drives. A number of reference frames have been proposed over the years [1], of which the stationary and synchronous reference frames are most commonly used. The transformation of variables between the two frames is presented below.

High-Power Converters and ac Drives. By Bin Wu
© 2006 The Institute of Electrical and Electronics Engineers, Inc.

14.2.1 *abc/dq* Frame Transformation

The transformation of the three-phase (*abc*-axis) variables of an induction motor to the equivalent two-phase (*dq*-axis) variables can be performed by

$$\begin{bmatrix} x_d \\ x_q \end{bmatrix} = \frac{2}{3} \begin{bmatrix} \cos\theta & \cos(\theta - 2\pi/3) & \cos(\theta - 4\pi/3) \\ -\sin\theta & -\sin(\theta - 2\pi/3) & -\sin(\theta - 4\pi/3) \end{bmatrix} \cdot \begin{bmatrix} x_a \\ x_b \\ x_c \end{bmatrix} \quad (14.2\text{-}1)$$

where x represents either current, voltage, or flux linkage, and θ is the angular displacement between the *a*-axis and *d*-axis of the three-phase and two-phase reference frames as shown in Fig. 14.2-1. The three-phase variables, x_a, x_b and x_c, are in the **stationary reference frame** which does not rotate in space whereas the two-phase variables, x_d and x_q, are in the **synchronous reference frame** whose direct (*d*) and quadrature (*q*) axes rotate in space at the synchronous speed ω_e. Note that ω_e is the angular electrical (not mechanical) speed of the rotating magnetic field of the motor, given by

$$\omega_e = 2\pi f_s \quad (14.2\text{-}2)$$

where f_s is the frequency of the stator variables. The angle θ can be found from

$$\theta(t) = \int_0^t \omega_e(t)dt + \theta_0 \quad (14.2\text{-}3)$$

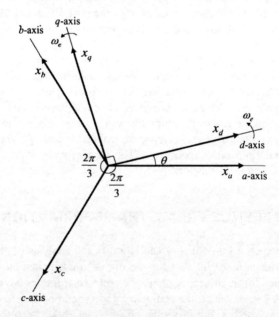

Figure 14.2-1 Variables in three-phase (*abc*) stationary frame and two-phase (*dq*) synchronous frame.

Note that the transformation equation of (14.2-1) is valid only for a three-phase balanced system, in which

$$x_a + x_b + x_c = 0 \quad (14.2\text{-}4)$$

Similarly, the two-phase variables in the synchronous frame can be transformed back to the three-phase stationary frame by

$$\begin{bmatrix} x_a \\ x_b \\ x_c \end{bmatrix} = \begin{bmatrix} \cos\theta & -\sin\theta \\ \cos(\theta - 2\pi/3) & -\sin(\theta - 2\pi/3) \\ \cos(\theta - 4\pi/3) & -\sin(\theta - 4\pi/3) \end{bmatrix} \cdot \begin{bmatrix} x_d \\ x_q \end{bmatrix} \quad (14.2\text{-}5)$$

which is referred to as **dq/abc transformation**.

The relationship between a space vector and its phase variables is illustrated in Fig. 14.2-2a, where a current space vector $\vec{i}_s$ rotates at a certain speed ω in the stationary frame (refer to Chapter 6 for space vector definition). Its phase currents i_{as}, i_{bs}, and i_{cs} can be obtained by decomposing $\vec{i}_s$ onto their corresponding *abc*-axes. Since the three axes are stationary in space, each of the phase currents varies one cycle over time when $\vec{i}_s$ rotates one revolution in space. If the length (magnitude) and the rotating speed of $\vec{i}_s$ are constant, the waveforms of the phase currents over time are sinusoidal with a phase displacement of $2\pi/3$ between each other.

Figure 14.2-2b illustrates another case where the current vector $\vec{i}_s$ is in the *dq*-axis synchronous reference frame. Assuming that $\vec{i}_s$ rotates at the same speed as that of the *dq*-axis frame, the stator current angle ϕ, which is the angle between $\vec{i}_s$ and the *d*-axis, is constant. The resultant *dq*-axis current components, i_{ds} and i_{qs}, are of dc signals. As will be seen in the subsequent sections, this transformation can be

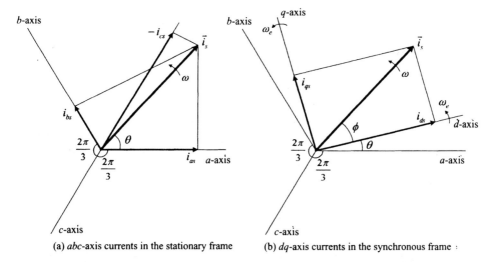

(a) *abc*-axis currents in the stationary frame (b) *dq*-axis currents in the synchronous frame

Figure 14.2-2 Decomposition of current vector $\vec{i}_s$.

utilized to simplify the simulation, design, and digital implementation of drive systems, where a three-phase ac signal can be effectively represented by a two-phase dc signal.

14.2.2 3/2 Stationary Transformation

With the rotating speed of the two-phase reference frame set at zero and its d-axis coincident with the a-axis of the three-phase frame ($\omega_e = 0$ and $\theta = 0$), both frames are stationary in space. The transformation of the three-phase variables to the two-phase variables can be obtained by setting θ in (14.2-1) to zero, from which

$$\begin{bmatrix} x_d \\ x_q \end{bmatrix} = \frac{2}{3} \begin{bmatrix} 1 & -1/2 & -1/2 \\ 0 & \sqrt{3}/2 & -\sqrt{3}/2 \end{bmatrix} \cdot \begin{bmatrix} x_a \\ x_b \\ x_c \end{bmatrix} \qquad (14.2\text{-}6)$$

The above transformation is referred to as **3/2 transformation** in this book. It is worth noting that the d-axis variable can be expressed as

$$x_d = \frac{2}{3}\left(x_a - \frac{1}{2}x_b - \frac{1}{2}x_c\right) = x_a \qquad (14.2\text{-}7)$$

which is equal to the a-axis variable x_a.

Similarly, the two-phase to three-phase stationary transformation, which is denoted **2/3 transformation**, can be performed by

$$\begin{bmatrix} x_a \\ x_b \\ x_c \end{bmatrix} = \begin{bmatrix} 1 & 0 \\ -1/2 & \sqrt{3}/2 \\ -1/2 & -\sqrt{3}/2 \end{bmatrix} \cdot \begin{bmatrix} x_d \\ x_q \end{bmatrix} \qquad (14.2\text{-}8)$$

14.3 INDUCTION MOTOR DYNAMIC MODELS

There are two commonly used dynamic models for the induction motor. One is based on space vector theory and the other is derived from dq-axis theory. The space vector model features compact mathematical expressions and concise space vector diagram, whereas the dq-axis model does not need to use complex numbers or variables. Both models are equally valid for the analysis of transient and steady-state performance of the induction motor. In what follows, the two models are presented and their relationship is revealed.

14.3.1 Space Vector Motor Model

It is assumed in the following analysis that the induction motor is three-phase symmetrical and its magnetic core is linear with a negligible core loss. The space vector model for an induction motor is generally composed of three sets of equations [2]. The first set is the voltage equations, given by

$$\vec{v}_s = R_s \vec{i}_s + p\vec{\lambda}_s + j\omega\vec{\lambda}_s$$
$$\vec{v}_r = R_r \vec{i}_r + p\vec{\lambda}_r + j(\omega - \omega_r)\vec{\lambda}_r \quad (14.3\text{-}1)$$

where $\vec{v}_s$ and $\vec{v}_r$ are the stator and rotor voltage vectors, respectively; $\vec{i}_s$ and $\vec{i}_r$ are the stator and rotor current vectors, respectively; $\vec{\lambda}_s$ and $\vec{\lambda}_r$ are the stator and rotor flux-linkage vectors, respectively; R_s, R_r are the stator and rotor winding resistance, respectively; ω is the rotating speed of an arbitrary reference frame; ω_r is the rotor angular speed (electrical); and p is the derivative operator ($p = d/dt$).

The terms $j\omega\vec{\lambda}_s$ and $j(\omega - \omega_r)\vec{\lambda}_s$ on the right-hand side of (14.3-1) are referred to as **speed voltages**, which are induced by the rotation of the reference frame.

The second set is the flux-linkage equations

$$\vec{\lambda}_s = L_s \vec{i}_s + L_m \vec{i}_r$$
$$\vec{\lambda}_r = L_r \vec{i}_r + L_m \vec{i}_s \quad (14.3\text{-}2)$$

where $L_s = L_{ls} + L_m$ represents the stator self-inductance; $L_r = L_{lr} + L_m$ represents the rotor self-inductance; L_{ls} and L_{lr} are the stator and rotor leakage inductances, respectively; and L_m is the magnetizing inductance. Note that all the rotor parameters and variables, such as R_r, L_{lr}, $\vec{i}_r$, and $\vec{\lambda}_r$, in the above questions are referred to the **stator side**.

The third set is the motion equation, given by

$$\frac{J}{P} p\omega_r = T_e - T_L$$
$$T_e = \frac{3P}{2} \text{Re}(j\vec{\lambda}_s \vec{i}_s^*) = -\frac{3P}{2} \text{Re}(j\vec{\lambda}_r \vec{i}_r^*) \quad (14.3\text{-}3)$$

where J is the total moment of inertia of the rotor and load, P is the number of pole pairs, T_L is the load torque, and T_e is the electromagnetic torque.

The above equations constitute the space vector model of the induction motor whose schematic representation is given in Fig. 14.3-1. The motor model is in the arbitrary reference frame that rotates in space at the arbitrary speed of ω.

In the simulation and digital implementation of advanced control systems, the motor models in the synchronous and stationary reference frames are often employed. The synchronous-frame motor model can be easily obtained by setting the arbitrary speed ω in (14.3-1) to the synchronous speed ω_e. Figure 14.3-2a shows the equivalent circuit of a squirrel cage motor in the synchronous frame, where the rotor winding is shorted ($\vec{v}_r = 0$) and ω_{sl} is the angular slip frequency, given by

$$\omega_{sl} = \omega_e - \omega_r \quad (14.3\text{-}4)$$

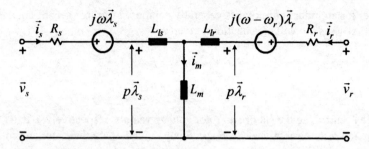

Figure 14.3-1 Space vector model of an induction motor in the arbitrary reference frame.

To obtain the model in the stationary (stator) frame, we can set the arbitrary speed ω of the rotating reference frame to zero. The resultant equivalent circuit is shown in Fig. 14.3-2b.

14.3.2 *dq*-Axis Motor Model

The induction motor *dq*-axis model can be derived using three-phase circuit theory and then transformed into the two-phase (*dq*-axis) frame [1]. Alternatively, it can

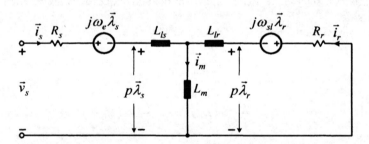

(a) Motor model in the synchronous frame

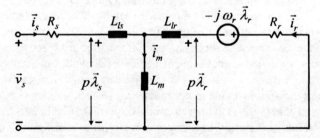

(b) Motor model in the stationary (stator) frame

Figure 14.3-2 Space vector models for a squirrel cage induction motor.

also be obtained by decomposing the space vectors in the space vector motor model into the d- and q-axis components [2], that is,

$$\begin{aligned} \vec{v}_s &= v_{ds} + jv_{qs}; & \vec{i}_s &= i_{ds} + ji_{qs}; & \vec{\lambda}_s &= \lambda_{ds} + j\lambda_{qs} \\ \vec{v}_r &= v_{dr} + jv_{qr}; & \vec{i}_r &= i_{dr} + ji_{qr}; & \vec{\lambda}_r &= \lambda_{dr} + j\lambda_{qr} \end{aligned} \quad (14.3\text{-}5)$$

Substituting (14.3-5) to (14.3-1), the dq-axis voltage equations for the induction motor can be obtained:

$$\begin{aligned} v_{ds} &= R_s i_{ds} + p\lambda_{ds} - \omega\lambda_{qs} \\ v_{qs} &= R_s i_{qs} + p\lambda_{qs} + \omega\lambda_{ds} \\ v_{dr} &= R_r i_{dr} + p\lambda_{dr} - (\omega - \omega_r)\lambda_{dr} \\ v_{qr} &= R_r i_{qr} + p\lambda_{qr} + (\omega - \omega_r)\lambda_{dr} \end{aligned} \quad (14.3\text{-}6)$$

where the stator and rotor flux-linkages can be calculated by

$$\begin{aligned} \lambda_{ds} &= L_{ls} i_{ds} + L_m(i_{ds} + i_{dr}) \\ \lambda_{qs} &= L_{ls} i_{qs} + L_m(i_{qs} + i_{qr}) \\ \lambda_{dr} &= L_{lr} i_{dr} + L_m(i_{ds} + i_{dr}) \\ \lambda_{qr} &= L_{lr} i_{qr} + L_m(i_{qs} + i_{qr}) \end{aligned} \quad (14.3\text{-}7)$$

The electromagnetic torque can be expressed in a number of ways. Some of the commonly used expressions are

$$T_e = \begin{cases} \dfrac{3P}{2}(i_{qs}\lambda_{ds} - i_{ds}\lambda_{qs}) \\ \dfrac{3PL_m}{2}(i_{qs}i_{dr} - i_{ds}i_{qr}) \\ \dfrac{3PL_m}{2L_r}(i_{qs}\lambda_{dr} - i_{ds}\lambda_{qr}) \end{cases} \quad (14.3\text{-}8)$$

Equations (14.3-6) to (14.3-8) together with the motion equation of (14.3-3) represent the dq-axis model of the induction motor, whose equivalent circuit is shown in Fig. 14.3-3.

14.3.3 Induction Motor Transient Characteristics

It is instructive to study transient characteristics of the induction motor during free acceleration using the dynamic motor models. The motor under investigation is a low-power squirrel cage motor with the following parameters: $V_{LL} = 208$ V, 60

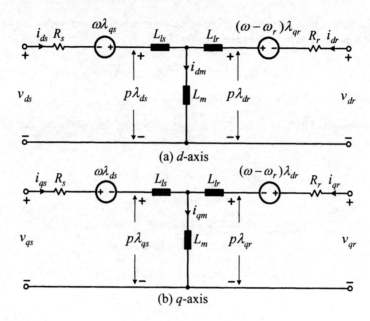

Figure 14.3-3 Induction motor dq-axis model in the arbitrary reference frame.

Hz, $Z_{base} = 15.4\ \Omega$, $R_s = 0.068$ pu, $R_r = 0.045$ pu, $L_{ls} = L_{lr} = 0.058$ pu, $L_m = 1.95$ pu, $P = 1$, and $J = 0.02$ kg·m². Figure 14.3-4 shows the block diagram for computer simulation with the motor model in the stationary frame ($\omega = 0$). The three-phase supply voltages v_{as}, v_{bs}, and v_{cs} are transformed to the dq-axis stator voltages v_{ds} and v_{qs} by the 3/2 transformation. The simulated dq-axis stator currents i_{ds} and i_{qs} are then converted to the three-phase currents i_{as}, i_{bs}, and i_{cs} by the 2/3 transformation.

Figure 14.3-5a shows simulated transient waveforms of the stator current i_{as} and rotor speed n_r during motor free acceleration (the motor starts under the rated volt-

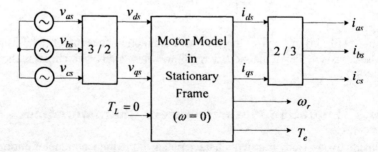

Figure 14.3-4 Block diagram for simulation of motor free acceleration using the stationary-frame motor model.

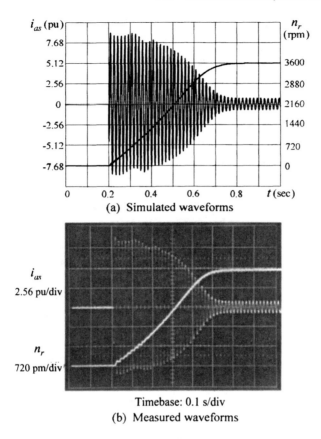

Figure 14.3-5 Waveforms of an induction motor during free acceleration.

age and frequency without mechanical load). The rotor speed n_r in rpm relates to the angular electrical rotor speed ω_r by

$$n_r = \frac{30}{\pi P}\omega_r \qquad (14.3\text{-}9)$$

The peak starting current is approximately 8.4 pu, which represents the rms starting current of 5.9 pu. The starting time is around 0.5 s due to the low moment of inertia and high starting current. The measured waveforms during the free acceleration are given in Fig. 14.3-5b, which match with the simulated results very well.

Figure 14.3-6 shows simulated and measured transient waveforms of the motor during a three-phase fault. The motor operates near its synchronous speed when its three-phase terminals are shorted. The maximum peak stator current is close to that during the free acceleration. The measured waveforms correlate closely with the simulated ones.

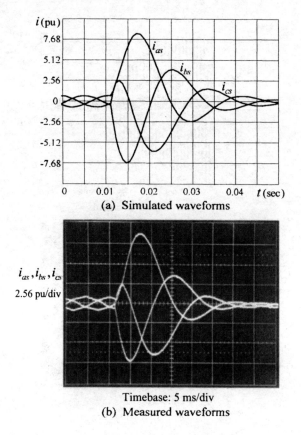

Figure 14.3-6 Stator current waveforms of an induction motor during a three-phase fault.

Figure 14.3-7 illustrates the simulation block diagram with the motor model in the synchronous frame. Using the *abc/dq* and *dq/abc* transformation blocks, the three-phase supply voltages v_{as}, v_{bs}, and v_{cs} in the stationary frame can be transformed to the *dq*-axis voltages v_{ds} and v_{qs} in the synchronous frame while the simulated *dq*-axis currents i_{ds} and i_{qs} in the synchronous frame can be converted to three-phase currents i_{as}, i_{bs}, and i_{cs} in the stationary frame. The angle θ in the transformation blocks can be obtained by the 3/2 transformation and $\tan^{-1}$ blocks shown in the figure or using Eq. (14.2-3) directly.

Figure 14.3-8a shows the simulated current waveforms for i_{as}, i_{ds} and i_{qs} during motor free acceleration. The waveform for i_{as} in the stationary frame is exactly the same as that in Fig. 14.3-5. The *dq*-axis currents i_{ds} and i_{qs} are in the synchronous frame, and therefore they are of dc signals in steady state. The amplitude of the stator current $\vec{i}_s$ can be obtained by $i_s = \sqrt{i_{ds}^2 + i_{qs}^2}$. The *dq*-axis voltages v_{ds} and v_{qs} in the synchronous frame are of a constant dc, and therefore they are not shown in the figure.

14.3 Induction Motor Dynamic Models

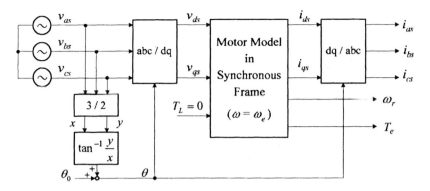

Figure 14.3-7 Block diagram for simulation of motor free acceleration using the synchronous-frame motor model.

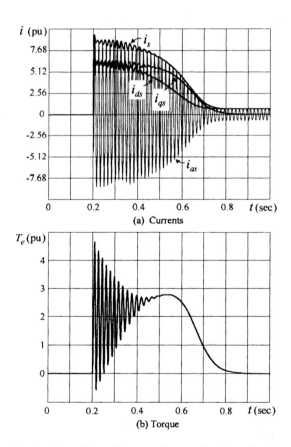

Figure 14.3-8 Simulated waveforms for motor free acceleration using the synchronous-frame motor model.

The stator current angle ϕ, which is the angle between $\vec{i}_s$ and the d-axis as shown in Fig. 14.2-2b, will affect waveforms for i_{ds} and i_{qs}. The waveforms in Fig. 14.3-8a are obtained with $\phi = 0$ by adjusting the initial angle θ_0 such that the d-axis of the synchronous frame is aligned with $\vec{i}_s$. This leads to $i_{qs} = 0$ and $i_{ds} = i_s$ when the motor is in steady-state operation. If the q-axis of the synchronous frame is aligned with $\vec{i}_s$ ($\phi = 90°$), the steady-state dq-axis currents are $i_{qs} = i_s$ and $i_{ds} = 0$. However, the waveform for i_s is not affected by ϕ.

The torque response during the motor free acceleration is shown in Fig. 14.3-8b. Although it is obtained with the motor model in the synchronous frame, the response remains the same for the motor model in any other reference frames.

14.4 PRINCIPLE OF FIELD-ORIENTED CONTROL (FOC)

14.4.1 Field Orientation

It is well known that the dc motor drive has an excellent dynamic performance. This is mainly due to the decoupled (separate) control of stator magnetic field and electromagnetic torque of the motor. The torque is developed by the interaction of two perpendicular magnetic fields. One field is generated by the field current i_f in the stator winding, and the other is produced by the armature (rotor) current i_a. The developed torque can be expressed as

$$T_e = K_a \lambda_f i_a \qquad (14.4-1)$$

where K_a is an armature constant and λ_f is the flux produced by i_f. In high-performance dc drives, λ_f is normally kept constant by keeping i_f constant, and thus the torque T_e is proportional to and can be directly controlled by i_a.

The field-oriented control, also known as **vector control**, for induction motor emulates the dc motor control. Using a proper field orientation, the stator current can be decomposed into a flux-producing component and a torque-producing component. These two components are then controlled separately.

The field orientation can be generally classified into stator flux, air-gap flux, and rotor flux orientations [3, 4]. Since the **rotor flux orientation** is extensively used in ac drives, this scheme is to be analyzed in detail. Its operating principle can be easily applied to the other two field orientation schemes.

The rotor flux orientation is achieved by aligning the d-axis of the synchronous reference frame with the rotor flux vector $\vec{\lambda}_r$ as shown in Fig. 14.4-1. The resultant d- and q-axis rotor flux components are

$$\lambda_{qr} = 0 \text{ and } \lambda_{dr} = \lambda_r \qquad (14.4-2)$$

where λ_r is the magnitude of $\vec{\lambda}_r$. Substituting (14.4-2) into the last equation of (14.3-8) yields

$$T_e = K_T \lambda_{dr} i_{qs} = K_T \lambda_r i_{qs} \qquad (14.4-3)$$

14.4 Principle of Field-Oriented Control (FOC)

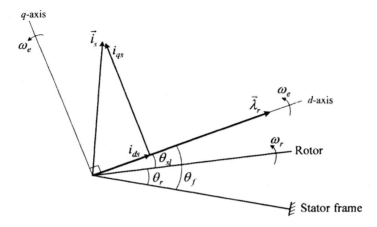

Figure 14.4-1 Rotor flux field orientation (the d-axis is aligned with $\vec{\lambda}_r$).

where $K_T = 3PL_m/2L_r$. Equation (14.4-3) indicates that with the rotor field orientation the torque expression for the induction motor is similar to that of a dc motor. If λ_r can be kept constant during the motor operation, the developed torque can be directly controlled by the q-axis stator current i_{qs}.

The stator current vector $\vec{i}_s$ in Fig. 14.4-1 can be resolved into two components along the dq-axes. The d-axis current i_{ds} is referred to as **flux-producing current** while the q-axis current i_{qs}, which is perpendicular to i_{ds}, is the **torque-producing current**. In the field-oriented control, i_{ds} is normally kept at its rated value while i_{qs} is controlled independently. With the decoupled control for i_{ds} and i_{qs}, a high-performance drive can be realized.

One of the key issues associated with the rotor flux-oriented control is to accurately determine the **rotor flux angle** θ_f for field orientation. Various schemes can be used to find θ_f. For instance, it can be calculated from measured stator voltages and currents, or it can be found from

$$\theta_f = \theta_r + \theta_{sl} \qquad (14.4\text{-}4)$$

where θ_r and θ_{sl} are the measured rotor position angle and calculated slip angle, respectively.

14.4.2 General Block Diagram of FOC

Depending on how the rotor flux angle θ_f is obtained, the control schemes can be classified into direct and indirect field-oriented controls. If θ_f is obtained by using flux-sensing devices embedded inside the motor or using measured motor terminal voltages and currents, the method is referred to as **direct field-oriented control**. If the rotor flux angle θ_f is obtained from detected rotor position angle θ_r and calculated slip angle θ_{sl} as shown in (14.4-4), this scheme is known as **indirect field-oriented control** [3, 4].

A general block diagram of an induction motor drive with rotor flux-oriented control is shown in Fig. 14.4-2. Since the essence of the FOC is the decoupled control of the rotor flux λ_r and electromagnetic torque T_e, these two variables are controlled separately. The torque reference T_e^* is generated by **Speed Controller** based on the reference speed ω_r^* and the detected or estimated rotor speed ω_r. The rotor flux reference λ_r^* is a function of ω_r^*. When the motor operates at or below its rated speed, λ_r^* is normally kept at its rated value. With the rated speed exceeded, λ_r^* should be weakened accordingly such that the stator voltage and the output power of the motor would not exceed their ratings.

The two references λ_r^* and T_e^* are then sent to **Flux/Torque Controller**, where they are compared with calculated rotor flux λ_r and torque T_e for a closed-loop control. The Flux/Torque Controller generates reference signals for the PWM block, which produces gate signals for the inverter to adjust its output voltage and frequency.

Based on the measured stator voltage and current variables and the motor model, the **Flux/Torque Calculator** calculates (1) the rotor flux angle θ_f for field orientation, (2) the rotor flux magnitude λ_r or the flux-producing current i_{ds}, (3) the electromagnetic torque T_e or the torque-producing current i_{qs}, and (4) the rotor speed ω_r. Depending on drive system requirements and the type of the FOC scheme employed, the rotor speed ω_r may also be directly detected by a digital speed sensor. It is worth noting that the Flux/Torque Calculator, also known as **Flux/Torque Observer** or Estimator in the literature, is the most important functional block in the FOC scheme.

14.5 DIRECT FIELD-ORIENTED CONTROL

14.5.1 System Block Diagram

Figure 14.5-1 shows a typical block diagram of direct field-oriented control for induction motor, where the rotor speed control is not shown for simplicity. There are

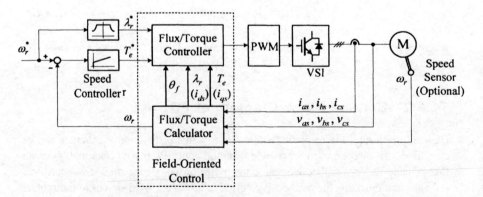

Figure 14.4-2 General block diagram of rotor flux FOC.

14.5 Direct Field-Oriented Control

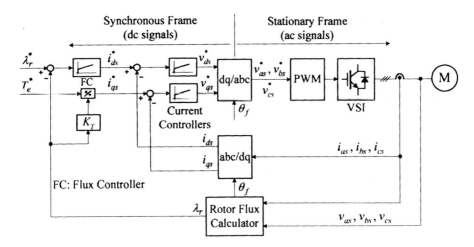

Figure 14.5-1 Direct field-oriented control with rotor flux orientation.

three feedback control loops, one for the rotor flux linkage λ_r, one for the d-axis (flux-producing) stator current i_{ds}, and another for the q-axis (torque-producing) stator current i_{qs}.

For the rotor flux control, the calculated λ_r is compared with its reference λ_r^* to generate the d-axis stator current reference i_{ds}^* through flux controller (FC). The q-axis stator current reference i_{qs}^* is generated according to the torque reference T_e^*. The feedback dq-axis stator currents i_{ds} and i_{qs} are compared with their references, and the errors are sent to current controllers to generate stator voltage references v_{ds}^* and v_{qs}^*. The dq-axis voltages in the synchronous frame are then transformed to the three-phase stator voltages v_{as}^*, v_{bs}^*, and v_{cs}^* in the stationary frame for the PWM block. Various PWM schemes can be used. If a carrier-based modulation scheme is employed, v_{as}^*, v_{bs}^*, and v_{cs}^* are the modulating signals that are compared with a triangular carrier wave to generate PWM gatings for the switching devices in the inverter.

As shown in Fig. 14.5-1, the rotor flux angle θ_f is used in the abc/dq and dq/abc transformation blocks for field orientation. The variables to the left of the transformation blocks are all dc signals in the synchronous frame, while those on the right side of the transformation blocks are all ac variables in the stationary frame.

14.5.2 Rotor Flux Calculator

Based on the stationary-frame motor model in Fig. 14.3-2b, the stator flux vector can be expressed as

$$\vec{\lambda}_s = \int (\vec{v}_s - R_s \vec{i}_s)\, dt \qquad (14.5\text{-}1)$$

The rotor flux vector can be found from the flux-linkage equations (14.3-2):

$$\vec{\lambda}_r = L_r \frac{\vec{\lambda}_s - L_s \vec{i}_s}{L_m} + L_m \vec{i}_s = \frac{L_r}{L_m}(\vec{\lambda}_s - \sigma L_s \vec{i}_s) \qquad (14.5\text{-}2)$$

where σ is the total leakage factor, defined by

$$\sigma = 1 - \frac{L_m^2}{L_s L_r} \qquad (14.5\text{-}3)$$

Decomposing the rotor flux $\vec{\lambda}_r$ into the d- and q-axis components, we have

$$\lambda_{dr} = \frac{L_r}{L_m}(\lambda_{ds} - \sigma L_s i_{ds})$$

$$\lambda_{qr} = \frac{L_r}{L_m}(\lambda_{qs} - \sigma L_s i_{qs}) \qquad (14.5\text{-}4)$$

from which the magnitude and angle of the rotor flux are

$$\lambda_r = \sqrt{\lambda_{dr}^2 + \lambda_{qr}^2}$$

$$\theta_f = \tan^{-1} \frac{\lambda_{qr}}{\lambda_{dr}} \qquad (14.5\text{-}5)$$

The following can be noted from (14.5-1) to (14.5-5):

- The rotor flux magnitude λ_r and its angle θ_f can be identified based on measured stator voltage $\vec{v}_s$, stator current $\vec{i}_s$ and motor parameters (L_s, L_r, L_m, and R_s).
- Since the stationary-frame motor model is used, all the variables such as λ_{dr}, λ_{qr}, i_{ds}, and i_{qs} (except λ_r and θ_f) are of ac signals. Neglecting the switching harmonics, they are sinusoidal in steady state.

Figure 14.5-2 shows the vector diagram for the rotor flux vector $\vec{\lambda}_r$ and stator current vector $\vec{i}_s$ used in the Rotor Flux Calculator. When the two vectors rotate one revolution in space, their dq-axis components λ_{dr}, λ_{qr}, i_{ds}, and i_{qs} in the stationary (stator) frame vary one cycle over time.

Figure 14.5-3 shows the block diagram for the digital implementation of Rotor Flux Calculator. Of the three stator voltages v_{as}, v_{bs}, and v_{cs}, only two need to be measured and the third one can be found from $v_{as} + v_{bs} + v_{cs} = 0$. To reduce the number of voltage sensors, the stator voltages can also be reconstructed by using the inverter switching function and measured dc voltage. The stator voltages and currents are then transformed to two-phase variables through 3/2 stationary transformation blocks. The other blocks are derived from equations (14.5-1) to (14.5-

14.5 Direct Field-Oriented Control

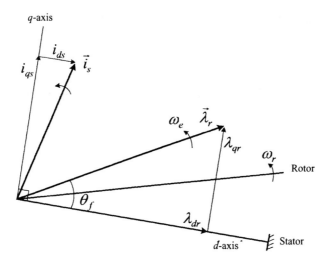

Figure 14.5-2 Vector diagram for $\vec{\lambda}_r$ and $\vec{i}_s$ used in rotor flux calculator.

5). The output of the Rotor Flux Calculator is the rotor flux angle θ_f and its amplitude λ_r.

14.5.3 Direct FOC with Current-Controlled VSI

The direct FOC scheme discussed in the previous sections can be simplified if a current-regulated voltage source inverter is used. Figure 14.5-4 shows the block diagram of such a system. The current-regulated VSI can make the inverter output currents i_{as}, i_{bs} and i_{cs} follow their references i_{as}^*, i_{bs}^*, and i_{cs}^* closely. Therefore,

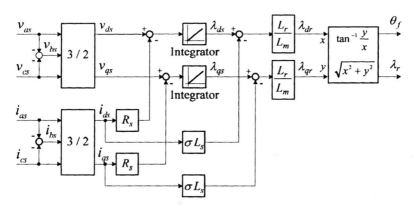

Figure 14.5-3 Block diagram for the rotor flux calculation.

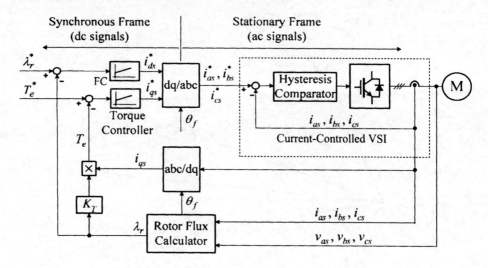

Figure 14.5-4 Direct FOC with a current regulated VSI.

there is no need to use the current controllers in Fig. 14.5-1. Furthermore, the PWM scheme for the current regulated VSI is simpler than other modulation techniques.

The operating principle of the current regulated VSI is illustrated in Fig. 14.5-5, where the control of only one of the three phase currents is shown. The inverter output current i_{as} is measured and compared with its reference i_{as}^*. Their difference Δi is sent to a hysteresis comparator. The output of the comparator x_H is either logic '1' or '0', based on which gate signals are generated for switches S_1 and S_4. The hysteresis comparator has a tolerance band of δ.

Assume that the reference current i_{as}^* is sinusoidal as illustrated in Fig. 14.5-5b. The inverter output current i_{as} is confined within the upper and lower band limits set by δ. Assuming that x_H becomes logic '1' at time instant t_1, S_1 is turned on and S_4 is off. The inverter terminal voltage v_{AN} is equal to the dc voltage V_d, causing i_{as} to increase. The rate of current rise is mainly determined by V_d, motor parameters and its back emf. When i_{as} reaches its upper band limit at t_2, x_H becomes logic '0', leading to the turn-off of S_1 and turn-on of S_4. The resultant v_{AN} is zero, causing i_{as} to decrease. When i_{as} hits the lower band limit at t_3, $x_H = $ '0', $v_{AN} = V_d$, and i_{as} starts to increase again. As a result, i_{as} is kept within the upper and lower band limits. To make i_{as} follow its reference i_{as}^* more closely with less switching harmonics, the bandwidth δ can be reduced. However, this is accomplished at the expense of an increased switching frequency.

For a given δ and V_d, the inverter switching frequency may vary with motor parameters. This is considered as a major drawback of the current-regulated inverter with the tolerance band control. When the inverter uses other modulation techniques such as SPWM and SVM, its switching frequency is set by the modulation scheme, independent of motor parameters.

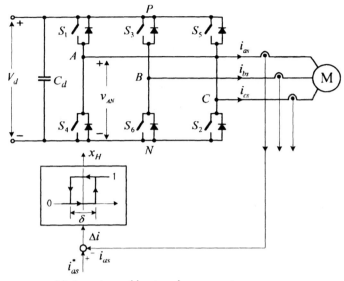

(a) Inverter and hysteresis comparator

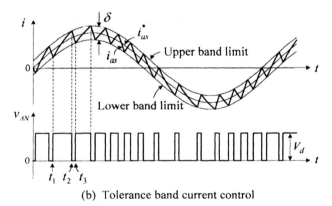

(b) Tolerance band current control

Figure 14.5-5 Current-regulated voltage source inverter.

Figure 14.5-6 shows the simulated waveforms for an induction motor drive using the direct FOC scheme. The block diagram of the drive system is shown Fig. 14.5-4, where the speed control loop is not shown. The motor nameplate data and parameters are listed in Table 14.5-1.

The tolerance band δ of the hysteresis comparator is adjusted such that inverter switching frequency is around 600 Hz. The rotor flux reference λ_r^* is set to its rated value of 8.35 Wb. The maximum torque is limited to its rated value of 7490 N·m during motor transients.

The motor initially operates at a rotor speed of $n_r = 200$ rpm. The speed reference n_r^* has a step increase from 200 rpm to the rated rotor speed of 1189 rmp at

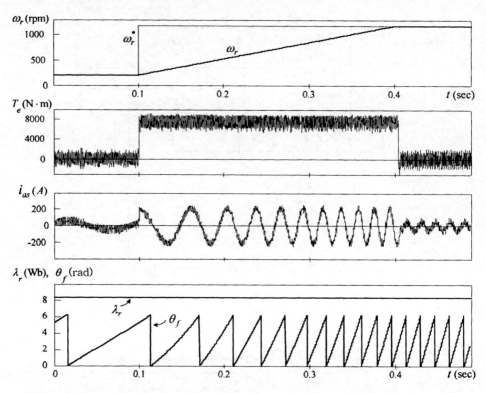

Figure 14.5-6 Simulated waveforms for an induction motor drive with direct FOC scheme.

time instant of $t = 0.1$ s. The motor accelerates under the no-load condition while its torque is limited to the rated value. The torque contains high ripples due to the low switching frequency. The stator current i_{as} rises to its rated value during the transient. When n_r reaches its reference of 1189 rpm around $t = 0.4$ s, the average T_e drops to zero, and i_{as} reduces to a value that corresponds to the magnetizing current of the motor. Due to the decoupled flux and torque control by the FOC scheme, the

Table 14.5-1 Motor Nameplate and Parameters

Motor Ratings		Motor Parameters	
Rated output power	1250 hp	Stator resistance, R_s	0.21 Ω
Rated line-to-line voltage	4160 V	Rotor resistance, R_r	0.146 Ω
Rated stator current	150 A	Stator leakage inductance, L_{ls}	5.2 mH
Rated speed	1189 rmp	Rotor leakage inductance, L_{lr}	5.2 mH
Rated torque	7490 N·m	Magnetizing inductance, L_m	155 mH
Rate stator flux linkage	9.0 Wb	Moment of inertia, J	22 kg·m²
Rated rotor flux linkage	8.35 Wb		

rotor flux magnitude λ_r is kept constant during transients. The rotor flux angle θ_f is also given in the figure.

14.6 INDIRECT FIELD-ORIENTED CONTROL

A digital rotor speed sensor is required for indirect field-oriented control schemes. The rotor flux angle θ_f for field orientation is obtained from the measured rotor speed and calculated slip angle based on motor parameters. A typical block diagram of the indirect FOC is shown in Fig. 14.6-1. Since the rotor speed ω_r is directly measured, the rotor flux angle θ_f can be found from

$$\theta_f = \int (\omega_r + \omega_{sl})\, dt \qquad (14.6\text{-}1)$$

where ω_{sl} is the angular slip frequency.

The slip frequency ω_{sl} can be derived from the synchronous-frame motor model of Fig. 14.3-2a, from which

$$p\vec{\lambda}_r = -R_r \vec{i}_r - j\omega_{sl}\vec{\lambda}_r \qquad (14.6\text{-}2)$$

Substituting the rotor current

$$\vec{i}_r = \frac{1}{L_r}(\vec{\lambda}_r - L_m \vec{i}_s) \qquad (14.6\text{-}3)$$

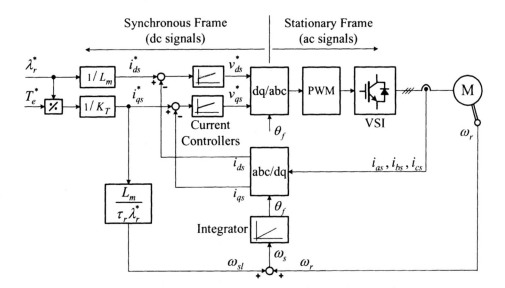

Figure 14.6-1 Indirect field-oriented control with rotor flux orientation.

into (14.6-2) yields

$$p\vec{\lambda}_r = -\frac{R_r}{L_r}(\vec{\lambda}_r - L_m \vec{i}_s) - j\omega_{sl}\vec{\lambda}_r \qquad (14.6\text{-}4)$$

from which

$$\vec{\lambda}_r(1 + \tau_r(p + j\omega_{sl})) = L_m \vec{i}_s \qquad (14.6\text{-}5)$$

where τ_r is the rotor time constant, defined by

$$\tau_r = L_r/R_r \qquad (14.6\text{-}6)$$

Decomposing (14.6-5) into the dq-axis components and taking into account the rotor flux orientation ($j\lambda_{qr} = 0$ and $\lambda_{dr} = \lambda_r$), we have

$$\begin{aligned} \lambda_r(1 + p\tau_r) &= L_m i_{ds} \\ \omega_{sl}\tau_r \lambda_r &= L_m i_{qs} \end{aligned} \qquad (14.6\text{-}7)$$

from which

$$\omega_{sl} = \frac{L_m}{\tau_r \lambda_r} i_{qs} \qquad (14.6\text{-}8)$$

As shown in Fig. 14.6-1, the rotor flux and torque are controlled by two feedback loops separately. Based on (14.6-7), the relationship between the rotor flux reference λ_r^* and d-axis current reference i_{ds}^* can be expressed as

$$i_{ds}^* = \frac{(1 + p\tau_r)}{L_m} \lambda_r^* \qquad (14.6\text{-}9)$$

Since λ_r^* is normally kept constant during operation ($p\lambda_r^* = 0$), (14.6-9) can be simplified to

$$i_{ds}^* = \frac{1}{L_m} \lambda_r^* \qquad (14.6\text{-}10)$$

The q-axis current reference i_{qs}^* can be obtained from the torque equation of (14.4-3):

$$i_{qs}^* = \frac{1}{K_T \lambda_r^*} T_e^* \qquad (14.6\text{-}11)$$

For a given λ_r^*, the torque-producing current i_{qs}^* is proportional to T_e^*.

14.7 FOC FOR CSI-FED DRIVES

In VSI-fed MV drives, both inverter output voltage and frequency can be controlled by its PWM scheme. However, this is not the case for the CSI-fed drive shown in Fig. 14.7-1, where the inverter output frequency is controlled by its PWM scheme whereas the inverter output current i_w is adjusted by dc current i_{dc} of the rectifier. In addition, the stator current i_s is not directly controlled by i_w due to the filter capacitor C_f. Therefore, additional measures are required to maintain field orientation in CSI drives.

Figure 14.7-2 shows the space vector diagram for the CSI drive with rotor flux oriented control. The d-axis of the synchronous reference frame is aligned with the rotor flux vector $\vec{\lambda}_r$. The stator flux vector $\vec{\lambda}_s$ leads the rotor flux $\vec{\lambda}_r$ by a small angle due to the leakage inductances. The stator voltage $\vec{v}_s$ is the sum of the speed voltage $j\omega_e\vec{\lambda}_s$ in Fig. 14.3-2a and the stator resistance voltage drop $R_s\vec{i}_s$. The stator current $\vec{i}_s$ lags $\vec{v}_s$ by θ_m, which is the motor power factor angle. The capacitor current $\vec{i}_c$ leads $\vec{v}_s$ by $\pi/2$. The inverter PWM current $\vec{i}_w$ is a vector sum of $\vec{i}_s$ and $\vec{i}_c$, and its angle with respect to $\vec{\lambda}_r$ is θ_w. The inverter firing angle is

$$\theta_{inv} = \theta_w + \theta_f \tag{14.7-1}$$

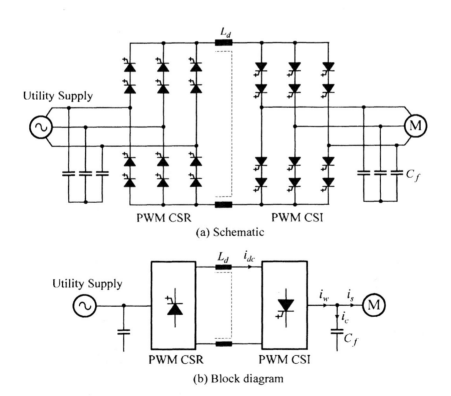

Figure 14.7-1 A PWM current source converter-based drive.

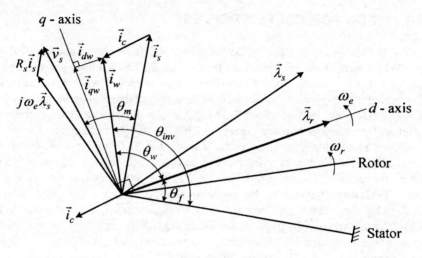

Figure 14.7-2 Vector diagram for a CSI-fed drive with rotor flux orientation.

where θ_f is the rotor flux angle for field orientation. When the drive operates in steady state, θ_w is constant since both $\vec{i}_w$ and $\vec{\lambda}_r$ are in the synchronous reference frame while θ_f and θ_{inv} vary periodically between zero and 2π.

Figure 14.7-3 shows a simplified block diagram for the CSI-fed drive with direct field-oriented control. The FOC scheme is implemented with three feedback control loops, one for the rotor speed ω_r, one for the rotor flux λ_r, and another for the dc cur-

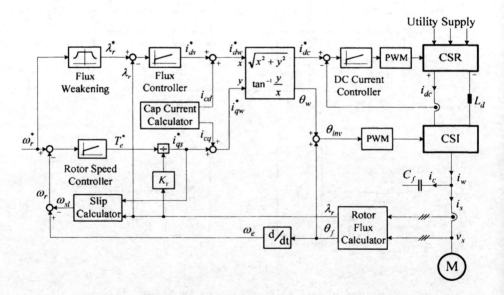

Figure 14.7-3 Simplified block diagram for a CSI-fed drive using direct FOC scheme.

rent i_{dc}. The rotor speed ω_r is obtained by $\omega_r = \omega_e - \omega_{sl}$, where $\omega_e = d\theta_f/dt$ and ω_{sl} can be calculated by (14.6-8).

The q-axis (torque-producing) stator current reference i_{qs}^* and d-axis (flux-producing) current reference i_{ds}^* are generated in the same manner as that shown in Fig. 14.5-1. The dq-axis inverter PWM current references can be expressed as

$$i_{dw}^* = i_{cd} + i_{ds}^*$$
$$i_{qw}^* = i_{cq} + i_{qs}^* \qquad (14.7\text{-}2)$$

where i_{cd} and i_{cq} are the dq-axis capacitor currents, given by

$$i_{cd} = (pv_{ds} - \omega_e v_{qs})C_f$$
$$i_{cq} = (pv_{qs} + \omega_e v_{ds})C_f \qquad (14.7\text{-}3)$$

The first term on the right-hand side of the equation represents capacitor transient current, and the second term is the steady-state current. To reduce the sensitivity and noise caused by the derivative terms (pv_{ds} and pv_{qs}), the effect of the capacitor transient response on the drive dynamic performance may be neglected. Equation (14.7-3) can then be simplified to

$$i_{cd} = -\omega_e v_{qs} C_f$$
$$i_{cq} = \omega_e v_{ds} C_f \qquad (14.7\text{-}4)$$

for use in the Capacitor Current Calculator in Fig. 14.7-3.

Since the magnitude of i_w is proportional to the dc current, the dc current reference can be found from

$$i_{dc}^* = \sqrt{(i_{dw}^*)^2 + (i_{qw}^*)^2} \qquad (14.7\text{-}5)$$

The inverter firing angle θ_{inv} is the sum of θ_f and θ_w, where θ_f can be obtained from the Rotor Flux Calculator in Fig. 14.5-3 and θ_w can be determined by

$$\theta_w = \tan^{-1}(i_{qw}^*/i_{dw}^*) \qquad (14.7\text{-}6)$$

Various PWM schemes, such as SHE, TPWM, and SVM presented in Chapters 10 and 11, can be employed for the PWM blocks in Fig. 14.7-3.

14.8 DIRECT TORQUE CONTROL

Direct torque control (DTC) is one of the advanced control schemes for ac drives [5–7]. It is characterized by simple control algorithm, easy digital implementation and robust operation [8, 9]. In this section, the principle of DTC scheme is introduced and simulation results are provided.

14.8.1 Principle of Direct Torque Control

The electromagnetic torque developed by an induction motor can be expressed in a number of ways, one of which is

$$T_e = \frac{3P}{2} \frac{L_m}{\sigma L_s L_r} \lambda_s \lambda_r \sin \theta_T \qquad (14.8\text{-}1)$$

where θ_T is the angle between the stator flux vector $\vec{\lambda}_s$ and rotor flux vector $\vec{\lambda}_r$, often known as **torque angle**. This equation indicates that T_e can be directly controlled by θ_T.

The main variable to be controlled in the DTC scheme is the stator flux vector $\vec{\lambda}_s$. Referring to the induction motor model of Fig. 14.3-2b, $\vec{\lambda}_s$ relates to the stator voltage vector $\vec{v}_s$ by

$$p\vec{\lambda}_s = \vec{v}_s - R_s \vec{i}_s \qquad (14.8\text{-}2)$$

The equation shows that the derivative of $\vec{\lambda}_s$ reacts instantly to changes in $\vec{v}_s$. As discussed in Chapter 6, the stator voltage $\vec{v}_s$, which is the inverter output voltage, can be controlled by the reference vector $\vec{V}_{ref}$ in the space vector modulation. Since $\vec{V}_{ref}$ is synthesized by the stationary voltage vectors of the inverter, a proper selection of the stationary vectors can make the magnitude and angle of $\vec{\lambda}_s$ adjustable.

Figure 14.8-1 shows the principle of direct torque control for a two-level VSI-fed induction motor drive. The dq-axis plane for the stator flux $\vec{\lambda}_s$ is divided into six sectors I to VI. The stator flux $\vec{\lambda}_s$ in the figure falls into sector I, and its angle θ_s is

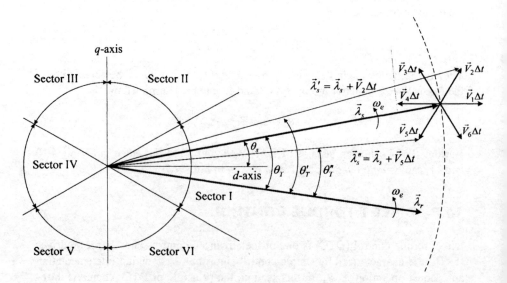

Figure 14.8-1 Principle of direct torque control.

referenced to the d-axis of the stationary reference frame. The rotor flux vector $\vec{\lambda}_r$ lags $\vec{\lambda}_s$ by θ_T.

Let's now examine the impact of the stationary voltage vectors $\vec{V}_0$ to $\vec{V}_6$ on $\vec{\lambda}_s$ and θ_T. Assume that $\vec{\lambda}_s$ and θ_T in Fig. 14.8-1 are the initial stator flux vector and torque angle. When voltage vector $\vec{V}_2$ is selected, the stator flux vector will become $\vec{\lambda}'_s = \vec{\lambda}_s + \vec{V}_2 \Delta t$ after a short time interval Δt, leading to an increase in flux magnitude ($\lambda'_s > \lambda_s$) and torque angle ($\theta'_T > \theta_T$). If voltage vector $\vec{V}_5$ is selected, $\vec{\lambda}_s$ will change to $\vec{\lambda}'' = \vec{\lambda}_s + \vec{V}_5 \Delta t$, causing a decrease in flux magnitude ($\lambda''_s < \lambda_s$) and torque angle ($\theta''_T < \theta_T$). Similarly, the selection of $\vec{V}_3$ and $\vec{V}_6$ can make one variable increase and the other decrease. Therefore, λ_s and θ_T can be controlled separately by proper selection of the inverter voltage vectors.

Note that the changes in $\vec{v}_s$ have much less impact on $\vec{\lambda}_r$ for a short time interval Δt due to the large rotor time constant. Therefore, it is assumed in the above analysis that the rotor flux vector $\vec{\lambda}_r$ is kept constant during Δt.

14.8.2 Switching Logic

Figure 14.8-2 shows a typical block diagram of a DTC-based induction motor drive, where the rotor speed feedback loop is not shown for simplicity. Similar to the FOC schemes, the stator flux and electromagnetic torque are controlled separately to achieve superior dynamic performance. The stator flux reference λ_s^* is compared with the calculated stator flux λ_s, and the error $\Delta\lambda_s$ is sent to Flux Comparator. The torque reference T_e^* is compared with the calculated torque T_e, and their difference ΔT_e is the input of Torque Comparator. The output of the flux and torque comparators (x_λ and x_T) are sent to Switching Logic unit for proper selection of the voltage vectors (switching states) of the inverter.

Both flux and torque comparators are of a hysteresis (tolerance band) type, whose transfer characteristics are shown in Fig. 14.8-3. The flux comparator has

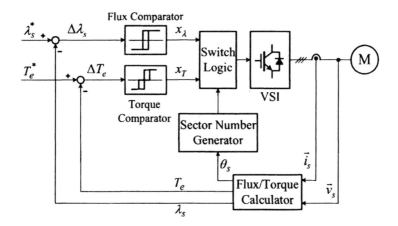

Figure 14.8-2 Block diagram of direct torque control scheme.

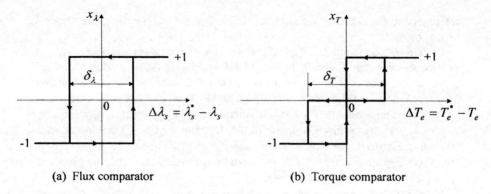

(a) Flux comparator (b) Torque comparator

Figure 14.8-3 Characteristics of hysteresis comparators.

two output levels ($x_\lambda = +1$ and -1) while the torque comparator has three output levels ($x_T = +1$, 0 and -1), where '+1' requests an increase in λ_s or θ_T, '−1' demands a decrease in λ_s or θ_T, and '0' signifies no changes. The tolerance band for the flux and torque comparators is δ_λ and δ_T, respectively.

Table 14.8-1 gives the switching logic for the stator flux reference $\vec{\lambda}_s^*$ rotating in the counterclockwise direction. The input variables are x_λ, x_T, and the sector number, and the output variables are the inverter voltage vectors. The output of the comparators decides which voltage vector should be selected. Assuming that $\vec{\lambda}_s^*$ is in sector I, the comparator output of $x_\lambda = x_T = +1$ signifies an increase in λ_s and T_e. Voltage vector $\vec{V}_2$ can then be selected from the table. This selection will make both λ_s and θ_T increase as shown in Fig. 14.8-1.

When the output of the torque comparator x_T is zero (no need to adjust T_e), zero vector $\vec{V}_0$ can be selected. The alternative use of the switching states [OOO] and

Table 14.8-1 Switching Logic for $\vec{\lambda}_s^*$ Rotating in the Counterclockwise Direction

Comparator Output		Sector					
x_λ	x_T	I	II	III	IV	V	VI
+1	+1	$\vec{V}_2$ [PPO]	$\vec{V}_3$ [OPO]	$\vec{V}_4$ [OPP]	$\vec{V}_5$ [OOP]	$\vec{V}_6$ [POP]	$\vec{V}_1$ [POO]
+1	0	$\vec{V}_0$ [PPP]	$\vec{V}_0$ [OOO]	$\vec{V}_0$ [PPP]	$\vec{V}_0$ [OOO]	$\vec{V}_0$ [PPP]	$\vec{V}_0$ [OOO]
+1	−1	$\vec{V}_6$ [POP]	$\vec{V}_1$ [POO]	$\vec{V}_2$ [PPO]	$\vec{V}_3$ [OPO]	$\vec{V}_4$ [OPP]	$\vec{V}_5$ [OOP]
−1	+1	$\vec{V}_3$ [OPO]	$\vec{V}_4$ [OPP]	$\vec{V}_5$ [OOP]	$\vec{V}_6$ [POP]	$\vec{V}_1$ [POO]	$\vec{V}_2$ [PPO]
−1	0	$\vec{V}_0$ [OOO]	$\vec{V}_0$ [PPP]	$\vec{V}_0$ [OOO]	$\vec{V}_0$ [PPP]	$\vec{V}_0$ [OOO]	$\vec{V}_0$ [PPP]
−1	−1	$\vec{V}_5$ [OOP]	$\vec{V}_6$ [POP]	$\vec{V}_1$ [POO]	$\vec{V}_2$ [PPO]	$\vec{V}_3$ [OPO]	$\vec{V}_4$ [OPP]

[PPP] for $\vec{V}_0$ in the switching table can help to reduce the device switching frequency. For instance, when x_T changes between '+1' and '0' or between '0' and '−1', the zero states in the table ensures that only two switches are involved during the transition, one being turned on and the other being turned off.

The operation of the direct torque control can be further explained by the stator flux trajectory diagram of Fig. 14.8-4. Assume that the reference vector $\vec{\lambda}_s^*$ rotates in the counterclockwise direction during rotor speed acceleration and the output of the torque comparator is $x_T = +1$. When $\vec{\lambda}_s$ reaches the outer band limit at point a in sector II, the output of the flux comparator x_λ becomes '−1', and vector $\vec{V}_4$ is selected from the switching table, which will cause a decrease in λ_s. When $\vec{\lambda}_s$ hits the inner band limit at point b, x_λ becomes +1, and vector $\vec{V}_3$ is selected, making an increase in λ_s. The trajectory of $\vec{\lambda}_s$ in the figure is not very smooth due to the wide width of the tolerance band δ_λ, which translates into a high stator flux ripple and low switching frequency. The quality of the stator flux waveform can be improved by reducing δ_λ, but it is achieved at the expense of an increase in the switching frequency.

The switching logic given in Table 14.8-1 is only valid for the motor rotating in the counterclockwise direction. When the motor operates in the clockwise direction, the switching logic in Table 14.8-2 can be used.

14.8.3 Stator Flux and Torque Calculation

The stator flux vector $\vec{\lambda}_s$ in the stationary frame can be expressed as

$$\vec{\lambda}_s = \lambda_{ds} + j\lambda_{qs}$$
$$= \int (v_{ds} - R_s i_{ds})\, dt + j\int (v_{qs} - R_s i_{qs})\, dt \qquad (14.8\text{-}3)$$

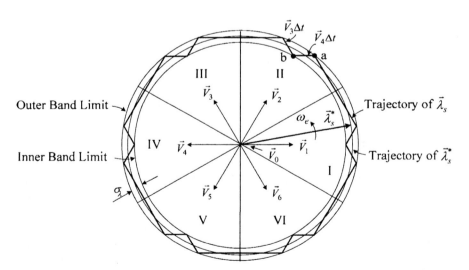

Figure 14.8-4 Trajectories of the stator flux $\vec{\lambda}_s$ and its reference $\vec{\lambda}_s^*$ with $x_T = 1$.

Table 14.8-2 Switching Logic for the Stator Flux Rotating in the Clockwise Direction

Hysteresis Comparator		Sector					
x_λ	x_T	I	II	III	IV	V	VI
+1	+1	$\vec{V}_6$ [POP]	$\vec{V}_5$ [OOP]	$\vec{V}_4$ [OPP]	$\vec{V}_3$ [OPO]	$\vec{V}_2$ [PPO]	$\vec{V}_1$ [POO]
+1	0	$\vec{V}_0$ [PPP]	$\vec{V}_0$ [OOO]	$\vec{V}_0$ [PPP]	$\vec{V}_0$ [OOO]	$\vec{V}_0$ [PPP]	$\vec{V}_0$ [OOO]
+1	−1	$\vec{V}_2$ [PPO]	$\vec{V}_1$ [POO]	$\vec{V}_6$ [POP]	$\vec{V}_5$ [OOP]	$\vec{V}_4$ [OPP]	$\vec{V}_3$ [OPO]
−1	+1	$\vec{V}_5$ [OOP]	$\vec{V}_4$ [OPP]	$\vec{V}_3$ [OPO]	$\vec{V}_2$ [PPO]	$\vec{V}_1$ [POO]	$\vec{V}_6$ [POP]
−1	0	$\vec{V}_0$ [OOO]	$\vec{V}_0$ [PPP]	$\vec{V}_0$ [OOO]	$\vec{V}_0$ [PPP]	$\vec{V}_0$ [OOO]	$\vec{V}_0$ [PPP]
−1	−1	$\vec{V}_3$ [OPO]	$\vec{V}_2$ [PPO]	$\vec{V}_1$ [POO]	$\vec{V}_6$ [POP]	$\vec{V}_5$ [OOP]	$\vec{V}_4$ [OPP]

from which its magnitude and angle are

$$\lambda_s = \sqrt{\lambda_{ds}^2 + \lambda_{qs}^2}$$
$$\theta_s = \tan^{-1}\left(\frac{\lambda_{qs}}{\lambda_{ds}}\right) \quad (14.8\text{-}4)$$

where v_{ds}, v_{qs}, i_{ds}, and i_{qs} are the measured stator voltages and currents. The developed electromagnetic torque can be calculated by

$$T_e = \frac{3P}{2}(i_{qs}\lambda_{ds} - i_{ds}\lambda_{qs}) \quad (14.8\text{-}5)$$

The above equations illustrate that the stator flux and developed torque can be obtained by using measured stator voltages and currents. The only motor parameter required in the calculations is the stator resistance R_s. This is in contrast to the direct rotor flux FOC schemes, where almost all the motor parameters are needed.

14.8.4 DTC Drive Simulation

Figure 14.8-5 shows the simulated waveforms for an induction motor drive using the DTC scheme given in Fig. 14.8-2. The rotor speed feedback loop, based on which the torque reference T_e^* is generated, is not shown for simplicity. The motor parameters used in the simulation are given in Table 14.5-1.

The tolerance bands δ_T and δ_λ for the torque and flux comparators are adjusted such that average switching frequency f_{sw} of the switching devices is around 800 Hz. The stator flux reference λ_s^* is set to its rated value of 9.0 Wb.

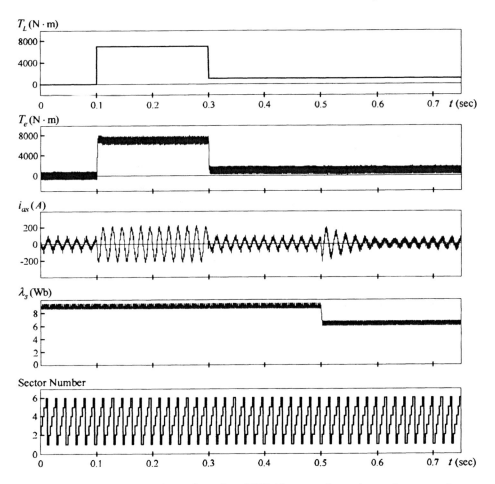

Figure 14.8-5 Simulated waveforms for a DTC drive operating at the rated rotor speed.

The motor operates at the rated speed of $n_r = 1189$ rpm under no-load conditions. Assuming that the load torque T_L is suddenly increased to its rated value of 7490 N·m at $t = 0.1$ s and then decreased to 1000 N·m at $t = 0.3$ s, the generated torque T_e responds quickly. The torque ripple is set by the torque tolerance band δ_T. The stator current i_{as} varies with T_e accordingly.

Since the stator flux λ_s and the motor torque T_e are controlled independently, λ_s is kept constant during the sudden load torque changes. To demonstrate the effectiveness of the stator flux control, its reference λ_s^* has a step reduction from its rated value of 9.0 Wb to 6.3 Wb at $t = 0.5$ s. The stator flux λ_s responds quickly while the stator current i_{as} is adjusted accordingly to keep T_e constant. The sector number obtained from the sector number generator in Fig. 14.8-2 is also shown in the figure.

316 Chapter 14 Advanced Drive Control Schemes

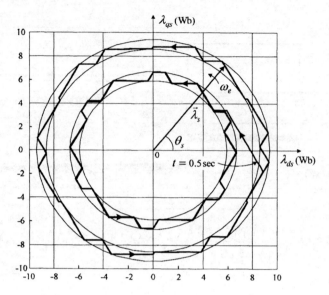

Figure 14.8-6 Trajectories of the stator flux λ_s in Fig. 14.8-5 for $0.35 \leq t \leq 0.75$ s.

Figure 14.8-6 shows the trajectories of the stator flux λ_s in Fig. 14.8-5 for $0.35 \leq t \leq 0.75$ s. The outer and inner trajectories correspond to the steady-state operation before and after the stator flux reduction at $t = 0.5$ s.

14.8.5 Comparison Between DTC and FOC Schemes

Based on the analysis given in the preceding sections, the features and drawbacks for the DTC and rotor flux FOC schemes are summarized in Table 14.8-3.

14.9 SUMMARY

The field-oriented control (FOC) and direct torque control (DTC) schemes for high-performance MV drives are presented in this chapter. There exist a variety of field oriented control schemes with many different variations and approaches. To make the subject matter easy to understand, this chapter focuses on the rotor flux oriented control schemes. The other reason for selecting such a scheme for detailed discussion lies in its simplicity and wide acceptance in the drive industry. The control algorithms for the direct and indirect rotor flux orientation are elaborated. In addition, the FOC for CSI-fed MV drives is introduced. The operating principle of the DTC scheme is discussed in detail, based on which a comparison between the two advanced schemes is provided.

The implementation of the FOC and DTC schemes requires accurate information on motor parameters. However, the motor parameters may vary with the operating

Table 14.8-3 Comparison Between DTC and FOC Schemes

Comparison	DTC	FOC
Field orientation (Reference frame transformation)	Not required	Required
Control scheme	Simple	Complex
Stator current control	No	Yes
Motor parameters required	R_s	$R_s, L_{ls}, L_{lr}, L_m,$ and R_r
Sensitivity to motor parameter variations	Not very sensitive	Sensitive
PWM scheme	Hysteresis band	Carrier-based, SVM, or hysteresis band
Switching behavior	Variable	Defined (for carrier-based and SVM)

conditions, such as rotor temperature rise and magnetic core saturation. The issues concerning the parameter sensitivity and on-line motor parameter tuning are out of the scope of this book, and therefore they are not addressed in this chapter.

REFERENCES

1. P. C. Krause, O. Wasynczuk, and S. D. Sudhoff, *Analysis of Electric Machines and Drive Systems*, 2nd edition, Wiley-IEEE Press, New York, 2002.
2. I. Boldear and S. A. Nasar, *Electric Drives*, CRC Press, Boca Raton, FL, 1999.
3. D. W. Novotny and T. A. Lipo, *Vector Control and Dynamics of AC Drives*, Clarendon Press, New York, 1996.
4. P. Vas, *Sensorless Vector and Direct Torque Control*, Oxford University Press, New York, 1998.
5. J. N. Nash, Direct Torque Control, Induction Motor Vector Control Without an Encoder, *IEEE Transaction on Industry Applications*, Vol. 33, No. 2, pp. 333–341, 1997.
6. P. Pohjalainen and C. Stulz, Method and Apparatus for Direct Torque Control of a Three-Phase Machine, US Patent 5,734,249, 9 pages, March 1998.
7. S. Heikkila, Direct Torque Control Inverter Arrangement, US Patent 6,094,364, 13 pages, July 2000.
8. D. Casadei, F. Profumo, and A. Tani, FOC and DTC: Two Viable Schemes for Induction Motors Torque Control, *IEEE Transactions on Power Electronics*, Vol. 17, No. 5, pp. 779–787, 2002.
9. D. Telford, M. W. Dunnigan, and B. W. Williams, A Comparison of Vector Control and Direct Torque Control of an Induction Machine, *IEEE Power Electronics Specialists Conference (PESC)*, Vol. 4, pp. 421–426, 2000.

Abbreviations

ABB	Asea–Brown–Boveri
APOD	Alternative phase opposite disposition
CHB	Cascade H-bridge
CSI	Current source inverter
CSR	Current source rectifier
DF	Distortion factor
DPF	Displacement power factor
DTC	Direct torque control
ETO	Emitter turn-off thyristor
FC	Flux controller
FOC	Field oriented control
GCT	Gate communicated thyristor (also know as integrated gate commutated thyristor)
CTO	Gate turn-off thyristor
HPF	High pass filter
IEEE	Institute of Electrical and Electronics Engineers
IEGT	Injection enhanced gate transistor
IGBT	Insulated gate bipolar transistor
IPD	In-phase disposition
LCI	Load commutated inverter
LPF	Low pass filter
MCT	MOS controlled thyristor
MOSFET	Metal-oxide semiconductor field-effect transistor
MV	Medium voltage (2.3 KV to 13.8 KV)
NPC	Neutral point clamped
PCBB	Power converter building block
PF	Power factor (DF × DPF)
PI	Proportional and integral
PLL	Phase-locked loop
POD	Phase opposite disposition
PWM	Pulse width modulation
pu	Per unit
rms	Root mean square
rpm	Revolutions per minute

SCR	Silicon controlled rectifier (thyristor)
SHE	Selective harmonic elimination
SIT	Static induction thyristor
SM	Synchronous motor
SPWM	Sinusoidal pulse width modulation
SVM	Space–vector modulation
THD	Total harmonic distortion
TPWM	Trapezoidal pulse width modulation
VSI	Voltage source inverter

Appendix

Projects for Graduate-Level Courses

P.1 INTRODUCTION

To assist the student in understanding the course material and the instructor in evaluating student's performance, a number of simulation based projects can be assigned. The titles of these projects are as follows:

1. 12-Pulse Series-Type Diode Rectifier
2. 18-Pulse SCR Rectifier
3. Space-Vector Modulation Schemes for Two-level Voltage Source Inverter
4. Multilevel CHB Inverter with Carrier-Based Modulation Techniques
5. Three-Level NPC Inverter with Space-Vector Modulation
6. IPD and APOD Modulation Schemes for Multilevel Diode Clamped Inverters
7. Current Source Inverter with Space-Vector Modulation
8. TPWM and SHE Schemes for Current Source Inverters
9. Dual-bridge Current Source Rectifier
10. VSI Fed MV Drive with Common-Mode Voltage Mitigation
11. CSI Fed MV Drive with Common-Mode Voltage Mitigation
12. High-Performance Induction Motor Drive with Field-Oriented Control

It is suggested that five to six projects be selected for a one-semester graduate course. The detailed instruction for the projects and their answers will be included in Instructor's Manual. As an example, the instruction for Project 3 is given in the following text.

P.2 SAMPLE PROJECT

Project 3—Space Vector Modulation Schemes for Two-Level Voltage Source Inverter

- **Objectives**
  1) To understand the principle of space vector modulation; and
  2) To investigate the harmonic performance of the two-level voltage source inverter.
- **Suggested Simulation Software**
  Matlab/Simulink
- **System Spefications**

Inverter Topology:	Two-level voltage source inverter as shown in Fig. 6.1-1
Rated Inverter Output Power:	1 MVA
Rated Inverter Output Voltage:	4160 V (fundamental voltage, rms)
Rated Inverter Output Current:	138.8 A (fundamental, rms)
Rated dc Input Voltage:	Constant dc (to be determined)
Load:	RL load with a per-phase resistance of 0.9 pu and inductance of 0.31 pu, which gives the load impedance of 1.0 pu with a lagging power factor of 0.95. Note that the RL load is fixed for the inverter operating under various conditions.
Switching Devices:	Ideal switch (no power losses or forward voltage drops)

- **Project Requirements**

  Part A

  A.1 Determine the dc input voltage V_d that can produce a fundamental line-to-line voltage of 4160 V (rms) at the modulation index of $m_a = 1.0$.

  A.2 Determine the value of load resistance (Ω) and inductance (mH).

  Part B

  Develop a simulation program for the conventional SVM scheme using the seven-segment switching sequence given in Table 6.3-4. Run your simulation program for the tasks given in Table P.1.

  B.1 For each of the above tasks, draw waveforms (two cycles each) for the inverter line-to-line voltage V_{AB} (V) and inverter output current i_A (A).

  B.2 Plot the harmonic spectrum (0 to 60th harmonics) of v_{AB} normalized to the dc voltage V_d and i_A normalized to its rated fundamental component $I_{A1,RTD}$ (138.8 A). Find the THD of v_{AB} and i_A.

Table P.1 Simulation tasks for the conventional SVM scheme

Simulation Task	f_1 (Hz)	m_a	T_s (sec)
T.1	30	0.4	1/720
T.2	30	0.8	1/720
T.3	60	0.4	1/720
T.4	60	0.8	1/720

Table P.2 Simulation tasks for the modified SVM scheme

Simulation Task	f_1 (Hz)	m_a	T_s (sec)
T.5	30	0.8	1/720
T.6	60	0.8	1/720

B.3 Analyze your simulation results and draw conclusions.

Part C

Modify your simulation program developed in Part B such that even-order harmonics in v_{AB} can be eliminated. Use the switching sequence given in Table 6.3-5. Run your simulation program for the tasks given in Table P.2.

C.1 For each of the above tasks, draw the waveforms for v_{AB} and i_A.

C.2 Calculate harmonic spectrum and THD of v_{AB} and i_A.

C.3 Find harmonic content of v_{AB} versus m_a for the inverter operating at $f_1 = 60$ Hz and $T_s = 1/720$ sec.

C.4 Analyze your simulation results and draw conclusions.

- **Project Report**

The project report is composed of the following six parts:
1. Title page
2. Abstract
3. Introduction
4. Theory
5. Simulation results
6. Conclusions

P.3 ANSWERS TO SAMPLE PROJECT

A.1 $V_d = 5883$ V
A.2 $R = 16.4\ \Omega$ and $L = 14.2$ mH per phase

B.1 Simulated Waveforms

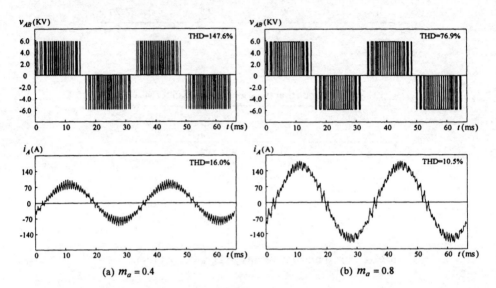

Figure P.1 Waveforms of v_{AB} and i_A at $f_1 = 30$ Hz.

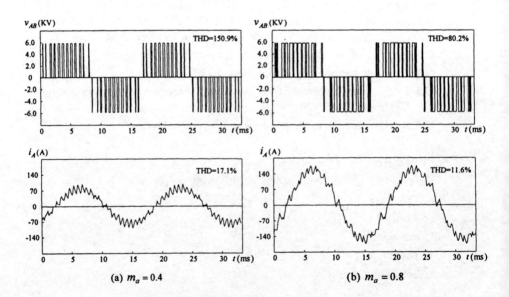

Figure P.2 Waveforms of v_{AB} and i_A at $f_1 = 60$ Hz.

B.2 Harmonic Spectrum and THD

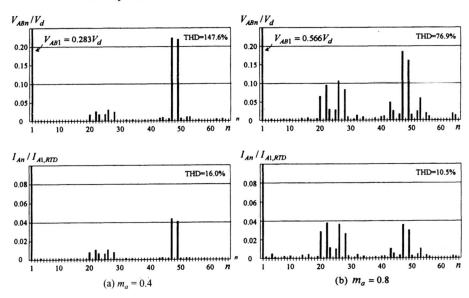

Figure P.3 Harmonic spectrum and THD of v_{AB} and i_A at $f_1 = 30$ Hz.

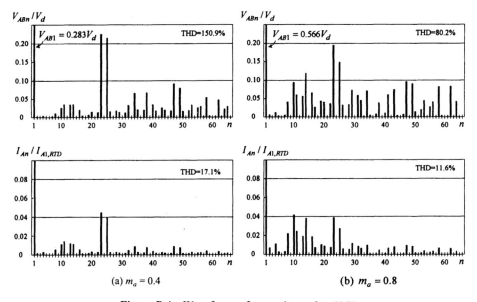

Figure P.4 Waveforms of v_{AB} and i_A at $f_1 = 60$ Hz.

B.3 Summary

- The waveform of v_{AB} is not half-wave symmetrical, i.e., $f(\omega t) \neq -f(\omega t + \pi)$. Therefore, it contains both even and odd order harmonics.
- The THD of i_A is much lower than that of v_{AB}. This is due to the filtering effect of the load inductance.
- The voltage and current harmonics appear in sidebands whose frequency is centered around the sampling frequency (720 Hz) and its multiples (such as 1440 Hz).
- The fundamental voltage V_{AB1} is proportional to the modulation index m_a.
- The THD of v_{AB} decreases with the increase of m_a, which is consistent with the THD curve in Fig. 6-3.7.
- The number of pulses N_p per half cycle of the inverter fundamental frequency does not affect the THD significantly. For example, the THD of v_{AB} in Figure P.1(a) with $N_p = 22$ is 147.6% in comparison to 150.9% in Figure P.2(a) where $N_p = 12$.
- The harmonic spectrum of v_{AB} in Figure P.4(b) is very close to the measured spectrum in Fig. 6.3-6.

C.1 Simulated Waveforms

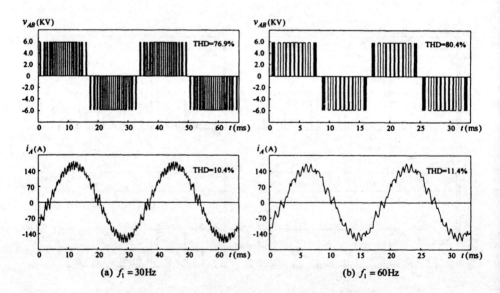

Figure P.5 Waveforms of v_{AB} and i_A at $m_a = 0.8$.

C.2 Harmonic Spectrum and THD

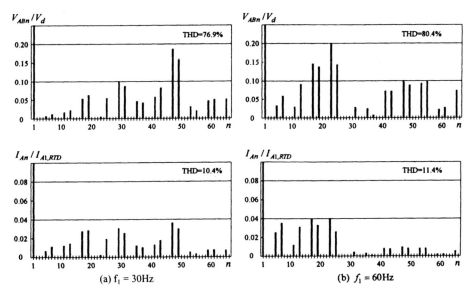

Figure P.6 Harmonic spectrum and THD of v_{AB} and i_A at $m_a = 0.8$.

C.3 Harmonic Content

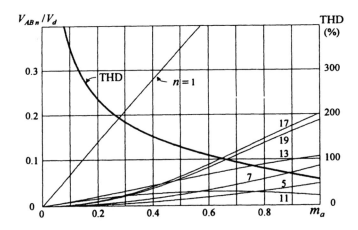

Figure P.3 Harmonic content ($f_1 = 60$ Hz and $T_s = 1/720$, no even order harmonics).

C.4 Summary

- The waveform of v_{AB} is of half-wave symmetry, i.e., $f(\omega t) = -f(\omega t + \pi)$. Therefore, it does not contain any even order harmonics.
- The THD of v_{AB} and i_A in Figure P.6 is almost identical to that given in Figures P.3(a) and P.4(b), which implies that the use of the modified SVM for even order harmonic elimination does not affect the THD profile of the inverter.
- The harmonic spectrum of v_{AB} in Figure P.6(b) is very close to the measured spectrum given in Fig. 6.3-10.

Index

12-pulse diode rectifier, 47,
   Line current THD, 46, 60
   Separate type, 57
   Series type, 49
12-pulse SCR rectifier, 74
   Effect of line inductance, 78
   Idealized, 75
   Power factor, 79
   THD, 77
18-pulse diode rectifier, 51
   Line current THD, 53, 61
   Separate type, 61
   Series type, 51
18-pulse SCR rectifier, 79
   Power factor, 79
   THD, 79
24-pulse diode rectifier, 54
   Separate type, 61
   Series type, 56
24-pulse SCR rectifier, 79

Active damping, 236, 240
Active overvoltage claming, 32,
Active switches, 126, 145
Active switching state, 200
Active vector, 101, 201
Alternative phase opposite disposition (APOD), 132
Amplitude modulation index, 95, 192, 222
Anode current, 19
APOD, *see* Alternative phase opposite disposition
Asymmetrical GCT, 24
Asynchronous PWM, 97

Bipolar PWM, 120
Blanking time, 97
Bypass operation, 200
Bypass pulse, 222

Carrier based PWM, 127, 170
Carrier wave, 95, 97
Cascaded H-bridge (CHB) inverter, 119, 261
CHB inverter, *see* Cascaded H-bridge inverter
Common mode voltage, 7, 256, 258, 277
Commutation, 145
CSI, *see* Current source inverter
CSR, *see* Current source rectifier
Current reference vector, 202
Current source inverter (CSI), 5, 189, 269
Current source rectifier (CSR), 219, 227, 269

Damping resistance, 238, 245
DC choke, 220
DC current balance control, 213
Delay time, 20, 24
Device switching frequency, 8, 120, 157, 164
di/dt, 21
Diode clamped inverter 143, 169
Direct field oriented control, 297, 301
Direct toque control (DTC), 309
   Stator flux calculation, 313
   Switching logic, 311
   Torque angle, 310
Discontinuous SVM 115, 167

330 Index

Displacement power factor, 44
Distortion factor, 44
DTC, *see* Direct toque control
Dual bridge CSR, 227, 275
dv/dt, 6, 21, 28, 148, 198, 256
Dwell time, 104, 149, 154, 203
Dynamic voltage equalization, 30
Dynamic voltage sharing, 148

Emitter turn off thyristor (ETO), 17
Even-order harmonic elimination, 111, 162

Fall time, 24, 26
Fiber-optic cable, 253
Field oriented control (FOC), 285, 296, 308
   Direct field oriented control, 297, 301
   Field orientation, 296
   Flux calculator, 298
   Flux controller, 299
   Flux observer, 298
   Flux-producing component, 297
   Indirect field oriented control, 298, 305
   Rotor flux angle, 297
   Rotor flux orientation, 296
   Slip frequency, 305
   Torque-producing component, 297
Field orientation, 296
Filter capacitor, 220, 235
Flux calculator, 298
Flux controller, 299
Flux observer, 298
Flux-producing component, 297
Flying capacitor inverter, 9, 183
FOC, *see* Field oriented control
Four-quadrant operation, 272
Frequency modulation index, 97

Gate commutated thyristor (GCT), 3, 190, 219
   Asymmetrical, 24
   Reverse conducting, 23
   Symmetrical, 23
Gate current, 20
Gate turn off thyristor (GTO), 3, 21, 189
GCT, *see* Gate commutated thyristor
GTO, *see* Gate turn off thyristor

Half-wave symmetry, 46,

Half-wave symmetrical, 108
Harmonic cancellation, 88
Harmonic content, 160, 172, 181, 208, 229
H-bridge inverter, 119, 179
Hexagon, 101, 168,
Hysteresis comparator, 303, 312

IEEE 519-1992
IEGT, *see* Injection enhanced gate transistor
IGBT, *see* Insulated gate bipolar transistor
IGCT, *see* Integrated gate commutated thyristor
Indirect field oriented control, 298, 305
Induction motor, 288
   dq-axis model, 290
   Dynamic model, 288
   Space vector model, 288
   Transient characteristics, 291
Injection enhanced gate transistor (IEGT), 17
In-phase disposition, 132, 138, 180
Input power factor, 6
Insulated gate bipolar transistor (IGBT), 3, 26, 253
Integrated gate commutated thyristor (IGCT), 23
Inverter leg, 119
Inverter phase voltage, 124
Inverter switching frequency, 122, 130, 161, 183
Inverter terminal voltage, 97, 144
IPD, *see* In-phase disposition

Large vector, 149
LC resonance, 6, 7, 237, 272
LCI, *see* Load commutated inverter
   Parallel resonant mode, 238
   Series resonant mode, 237
Level shifted PWM, 131, 138, 184
Line current distortion, 5
Line inductance, 40, 48, 219
Load commutated inverter (LCI), 189, 215, 281
Load phase voltage, 124

Maximum average on-state current, 21
Maximum modulation index, 106, 204
Maximum repetitive perk off-state voltage, 21

Maximum repetitive perk reverse voltage, 21
Maximum rms on-state current, 21
MCT, *see* MOS controlled thyristor
Medium vector, 149
Medium voltage drive, 3, 253
    CHB inverter fed, 261
    Current source fed, 269, 307
    NPC inverter fed, 257
    Two-level VSI fed, 253
Modular structure, 255, 263
Modulating wave, 95
Modulation index, 106, 132, 140, 153, 204
    Amplitude modulation index, 95, 132
    Frequency modulation index, 97, 133
MOS controlled thyristor (MCT), 17
Motor derating, 7
Multipulse diode rectifier, 37
    Separate type, 38
    Series type, 37
MV drive, *see* Medium voltage drive

N+1 redundancy, 264, 272
Natural commutation, 30
Neutral current, 144
Neutral point, 144
Neutral point clamped (NPC) inverter, 143, 179, 257
Neutral point voltage 144
    Control, 164
    Deviation, 144, 155, 165
    Feedback control, 166
Newton-Raphson algorithm, 198
Non-characteristic harmonics, 97
NPC inverter, *see* Neutral point clamped inverter

Overmodulation, 99
Overvoltage claming, 32,

Passive frond end, 254
PCBB, *see* Power converter building block
Per-unit system, 45
Phase opposite disposition (POD), 132
Phase shifted PWM, 127, 138, 184
Phase shifting transformer, 83
    Harmonic cancellation, 88
POD, *see* Phase opposite disposition

Power converter building block (PCBB), 253
Power factor, 44, 47, 55, 61, 72, 79, 231,
    Control, 231, 234
    Displacement, 44
    Distortion factor, 44
Press pack, 19, 32
Pulse width modulation, 95, 97, 120
    Asynchronous, 97
    Bipolar, 120
    Level shifted PWM, 131, 138, 184
    Phase shifted PWM, 127, 138, 184
    Synchronous, 97
    Third harmonic injection, 99, 130
    Unipolar, 121
PWM, *see* Pulse width modulation

Redundant switching state, 103
Reference frame transformation, 285
    2/3 transformation, 288, 294
    3/2 transformation, 288, 294
    abc/dq transformation, 286, 294, 299
    dq/abc transformation, 287, 294, 299
Reference vector, 103, 149, 154, 202
Reverse conducting GCT, 24
Reverse recovery charge, 21
Reverse recovery current, 21
Reverse recovery time, 21
Rise time, 20, 24, 26
Rotor flux angle, 297

Sampling frequency, 107
Sampling period, 104, 203
SCR, *see* Silicon controlled rectifier
Sector, 102, 149
Selective harmonic elimination (SHE), 189, 194, 209, 220
Series connection, 8
Seven-segments, 107
Silicon controlled rectifier (SCR), 18
Sinusoidal pulse width modulation (SPWM), 95
SIT, 17
Six-pulse diode rectifier, 38
    Capacitive load, 40
    Continuous current operation, 43
    Discontinuous current operation, 40
Six-pulse SCR rectifiers, 65
    Idealized, 66

Six-pulse SCR rectifiers *(continued)*
   Effect of line inductance, 70
   Power factor, 69
   THD, 67
Slip frequency, 305
Small vector, 149
Snubber, 26, 253
   Capacitor, 31
   Turn-on snubber, 26
Space vector modulation (SVM), 101, 143, 148, 200, 210
   Active switching state, 200
   Active vector, 101, 201
   Discontinuous SVM 115, 167
   Dwell time, 104, 149, 154, 203
   Large vector, 149
   Maximum modulation index, 106, 204
   Medium vector, 149
   Modulation index, 106, 153, 204
   Reference vector, 103, 149, 154, 202
   Sampling period, 104, 203
   Sector, 102, 149
   Seven-segment, 107
   Small vector, 149
   Space vector, 101, 201
   Stationary vector, 103, 149, 202
   Switching sequence, 107, 113, 154, 176, 205
   Switching sequence design, 205
   Switching state, 101, 144, 200
   Volt-second balancing, 104, 149
   Zero switching state, 200
   Zero vector, 101, 201
Spectrum analysis, 108
SPWM, *see* Sinusoidal pulse width modulation
Staircase modulation, 139
Static induction thyristor (SIT), 17
Static voltage equalization, 29
Static voltage sharing, 148
Stationary vector, 103, 149, 202
Stator flux calculation, 313
Storage time, 21, 24
SVM, *see* Space vector modulation
Switching angle, 218, 221
Switching frequency, 8, 97, 193, 222
   Device switching frequency, 8, 120, 157, 164

Inverter switching frequency, 122, 130, 161, 183
Switching logic, 311,
Switching sequence, 107, 113, 154, 176, 205
Switching sequence design, 205
Switching state, 101, 144, 200
Symmetrical GCT, 24
Synchronous PWM, 97

Tail time, 21
THD, *see* Total harmonic distortion
Third harmonic injection, 99, 130
Torque angle, 310
Torque-producing component, 297
Total harmonic distortion (THD), 43, 46, 57, 67, 79, 99, 135, 172, 224
TPWM, *see* Trapezoidal pulse width modulation
Trapezoidal pulse width modulation (TPWM), 189, 193, 270
Triple harmonics, 46
Two level voltage source inverter, 95
Turn-off delay time, 26,
Turn-off time, 20,
Turn-on delay time, 26,
Turn-on time, 21,
Turn-on transient, 30

Unequal dc voltage, 126
Unipolar PWM, 121

Voltage equalization, 29. 31
   Dynamic, 30
   Static, 29
Voltage source inverter, 5,
Voltage unbalance, 29
Volt-second balancing, 104, 149
VSI, *see* Voltage source inverter

Wave reflections, 6, 256

Zero sequence, 46
Zero switching state, 200
Zero vector, 101, 201

About the Author

Bin Wu is Professor of Electrical and Computer Engineering at Ryerson University and Ryerson Research Chair. He has published more than 100 papers, authored or coauthored 100 technical reports, and holds five U.S. patents with another four patents pending in power electronics and AC drives. Dr. Wu has closely collaborated with a number of manufacturing companies, including Rockwell Automation and Honeywell Aerospace Canada, assisting them in achieving technical and commercial success through research and new product development. He has received research funding totaling $3.5 million from government sources and the private sector.

Dr. Wu is an Associate Editor of *IEEE Transactions on Power Electronics* and the Chair of Industry Relations Committee of IEEE Canada. He is a Registered Professional Engineer in the Province of Ontario. He was also the founder of LEDAR—Laboratory for Electric Drive Applications and Research—the best of its kind in a Canadian university. His honors include the Gold Medal of the Governor General of Canada, the NSERC Synergy Award for Innovation, Premier Research Excellence Award, and Ryerson Sarwan Sahota Distinguished Scholar Award. Dr. Bin Wu received his Ph.D. degree in electrical and computer engineering from the University of Toronto.

Lightning Source UK Ltd.
Milton Keynes UK
UKOW020141040112

184699UK00001B/25/P